EXTRUDERS IN FOOD APPLICATIONS

EXTRUDERS IN FOOD APPLICATIONS

Edited by

Mian N. Riaz, Ph.D.
Head, Extrusion Technology Program
Food Protein Research and Development Center
Texas A&M University

CRC PRESS

Boca Raton   London   New York   Washington, D.C.

Library of Congress Cataloging-in-Publication Data

Main entry under title:
Extruders in Food Applications

Full Catalog record is available from the Library of Congress

This book contains information obtained from authentic and highly regarded sources. Reprinted material is quoted with permission, and sources are indicated. A wide variety of references are listed. Reasonable efforts have been made to publish reliable data and information, but the author and the publisher cannot assume responsibility for the validity of all materials or for the consequences of their use.

Neither this book nor any part may be reproduced or transmitted in any form or by any means, electronic or mechanical, including photocopying, microfilming, and recording, or by any information storage or retrieval system, without prior permission in writing from the publisher.

The consent of CRC Press LLC does not extend to copying for general distribution, for promotion, for creating new works, or for resale. Specific permission must be obtained in writing from CRC Press LLC for such copying.

Direct all inquiries to CRC Press LLC, 2000 N.W. Corporate Blvd., Boca Raton, Florida 33431.

Trademark Notice: Product or corporate names may be trademarks or registered trademarks, and are used only for identification and explanation, without intent to infringe.

Visit the CRC Press Web site at www.crcpress.com

© 2000 by CRC Press LLC
Originally Published by Technomic Publishing
First CRC Reprint 2002

No claim to original U.S. Government works
International Standard Book Number 1-56676-779-2
Library of Congress Card Number 99-68515
Printed in the United States of America   2 3 4 5 6 7 8 9 0
Printed on acid-free paper

Contents

Preface ix

List of Contributors xi

1. INTRODUCTION TO EXTRUDERS AND THEIR PRINCIPLES 1
MIAN N. RIAZ

Definitions of Extrusion 1
Functions of an Extruder 2
Advantages of Extrusion 3
Development of Extruders 4
Terminology 5
Types of Extruders 8
Single Screw Classification 8
Twin-Screw Extruders 14
Classification of Twin-Screw Extruders 14
Advantages of Twin-Screw Extruders 16
New Generation Extruders 18
Advantages of New Generation Extruders 18
Suggested Readings 20
References 21

2. SINGLE-SCREW EXTRUDERS 25
GALEN J. ROKEY

Raw Material Characteristics and Selection   26
Selection of Hardware Components   27
Processing Conditions   35
Applications   39
References   49

3. DRY EXTRUDERS 51
NABIL W. SAID

The Principle of Dry Extrusion   51
Classification of the Dry Extruder   52
Components of the Single-Screw Dry Extruder   52
The Application of Dry Extrusion   53
Nutritional Advantages of the Dry Extrusion Process   55
References   61

4. INTERRUPTED-FLIGHT EXPANDERS-EXTRUDERS 63
MAURICE A. WILLIAMS

Background of Interrupted Flighting   65
Making Dog Food   69
Making Fish Feed   70
Making Full-Fat Soy   72
Extrusion Before Solvent Extraction   73
Slotted-Wall Expanders   74
Extrusion Before Crushing   75
Extruder Drying of Synthetic Rubber   76
Summary   78
References   78

5. TWIN-SCREW EXTRUDERS 81
GORDON R. HUBER

Introduction   81
Background   82
Extrusion Cooking   82
Advantages of Extrusion   83
Past Concerns   84
Extruder Classification   85
Process Description   86
Description of Individual Components   88
Extrusion Processing Zones   91

Twin-Screw Drive Design   94
Screw and Barrel Design for Single-Screw Extruders   96
Screw and Barrel Design for Twin-Screw Extruders   97
Screw Design   98
Kneading Elements   99
Reverse Pitch   101
Conical Elements   103
Design Limits   106
Extrusion Processing Variables   108
Mass and Energy Balance   112
Conclusion   113
References   113

6. PRECONDITIONING 115
BRADLEY S. STRAHM

Benefits of Preconditioning   117
Preconditioning Hardware   119
Preconditioner Operations   122
References   126

7. CHEMICAL AND NUTRITIONAL CHANGES IN FOOD DURING EXTRUSION 127
MARY ELLEN CAMIRE

Critical Factors   128
Starch   129
Dietary Fiber   131
Protein   134
Lipids   135
Vitamins   137
Minerals   138
Phytochemicals   139
Natural Toxins   140
Flavors   142
Future Directions   142
References   142

8. PRACTICAL CONSIDERATIONS IN EXTRUSION PROCESSING 149
MIAN N. RIAZ

Which Extruder to Purchase   149
When to Buy a Twin-Screw Extruder   151

Common Extrusion Problems and Their Solutions   151
Start-Up Sequence for a Typical Extruder   157
Extrusion "Rules of Thumb"   160
References   165

9. EXTRUDERS IN THE FOOD INDUSTRY 167
ERIC SEVATSON and GORDON R. HUBER

Introduction   167
History and Uses of Extruders in the Food Industry   168
Textured Vegetable Protein (TVP) Production   171
Ready-to-Eat Breakfast Cereal Production   181
Direct Expanded (DX) and Third Generation (3G) Snacks   193
References   204

APPENDIX 205
Mass and Energy Evaluation in Extrusion Systems   205
Nomenclature   216
Useful Conversion Factors   219

Index   221

Preface

EXTRUSION processing in foods and feeds has become very popular. The subject of extrusion cooking is now of major importance in food and feed processing. Because extruders are being applied in so many diverse operations, they are increasingly regarded as a versatile process. Most new industries are installing extruders rather than the traditional processing systems.

No text currently exists on different types of extruders available for the food and feed industries. There are several different types of extruders available in the market, which makes it very difficult to select the proper type of extruder. This book will give insight into the four different types of extruders that are presently being used in the food and feed markets.

This book is written to summarize some of the fundamentals to be considered in the application of extrusion technology in the food and feed industries. This text is an excellent starting point for students and other professionals who are in food or feed extrusion. It brings together in-depth knowledge of extrusion cooking technology and practical experience in the application of this technology to the food and feed industry. There is a wealth of information about single-screw extruders, dry extruders, interrupted-flight extruder-expanders, and twin-screw extruders. Also discussed is the effect of preconditioning on the raw material and the effect of extrusion on the nutrients of products. What happens to food nutrients during extrusion cooking, some practical considerations in extrusion processing, and what kind of food can be processed by extrusion cooking are also covered. This book is a valuable source for the technical and practical applications of extrusion technology and will be useful for the selection of the proper equipment for this technology. This book is the result of practical experience in extrusion technology. Most of the contributors to this book have at least 15 to 20 years of practical experience in extrusion.

I believe this book will serve as a source of information to all who are involved with food and feed extrusion. For the person who is new in this area, this book will serve as a guide for understanding and properly selecting an extruder. I owe a large debt of gratitude to a number of individuals who provided information and inspiration. I especially wish to thank Anderson, Insta-Pro, and Wenger for providing technical information about their extruders.

List of Contributors

Mary Ellen Camire, Ph.D.
Associate Professor
Department of Food Science and
  Human Nutrition
University of Maine
5736 Holmes Hall
Orono, ME 04469

Gordon R. Huber, Director
New Concept Development
Wenger Manufacturing Co.
714 Main Street
Sabetha, KS 66534

Mian N. Riaz, Ph.D.
Head of the Extrusion
  Technology Program
Food Protein R&D Center
Texas A&M University System
College Station, TX 77843

Galen J. Rokey
Manager
Technical Center
Wenger Manufacturing Co.
714 Main Street
Sabetha, KS 66534

Nabil W. Said, Ph.D.
Director of Research &
  Development
Triple "F"/ Insta-Pro
  International
10104 Douglas Avenue
Des Moines, IA 50322

Eric Sevatson
Food Technologist
Wenger Manufacturing Co.
714 Main Street
Sabetha, KS 66534

Bradley S. Strahm
Process Development Engineer
Wenger Manufacturing Co.
714 Main Street
Sabetha, KS 66534

Maurice A. Williams
Director
Research & Development
Anderson International
   Corporation
6200 Harvard Avenue
Cleveland, OH 44105

CHAPTER 1

Introduction to Extruders and Their Principles

MIAN N. RIAZ

THE objectives of this chapter are to review the history of extruder development, introduce terminology, and review principles of extrusion processing that will be described in more detail by other authors. Discussions about specific machines in this book do not constitute endorsement or preference of products or services by The Texas A&M University System or its divisions.

DEFINITIONS OF EXTRUSION

Extrusion is simply the operation of shaping a plastic or dough-like material by forcing it through a restriction or die. Examples of hand operations for extruding foods include the rolling of noodles and pie crust doughs, finger-stuffing of chopped meats through animal horns into natural casings, pressing of soft foods through hand ricers to produce string-like particles, and cranking of hand-powered meat grinders. Mechanically powered extrusion devices include wire-cut cookie dough depositors, pasta presses, continuous mixing and scaling systems used in automated bakeries, pneumatic (batch) and continuous (pump) sausage stuffers, hamburger patty formers, and pellet mills used to prepare animal feeds. Rossen and Miller (1973) have offered the practical definition: "Food extrusion is a process in which a food material is forced to flow, under one or more varieties of conditions of mixing, heating and shear, through a die which is designed to form and/or puff-dry the ingredients."

A food extruder is a device that expedites the shaping and restructuring process for food ingredients. Extrusion is a highly versatile unit operation that can be applied to a variety of food processes. Extruders can be used to cook, form, mix, texturize, and shape food products under conditions that favor quality retention, high productivity, and low cost. The use of cooker extruders has been expanding rapidly in the food and feed industries over the past few years.

FUNCTIONS OF AN EXTRUDER

The conditions generated by the extruder permit the performance of many functions that allow it to be used for a wide range of food, feed, and industrial applications. Some of these functions are as follows:

Agglomeration: Ingredients can be compacted and agglomerated into discrete pieces with an extruder.

Degassing: Ingredients that contain gas pockets can be degassed by extrusion processing.

Dehydration: During normal extrusion processing, a moisture loss of 4–5% can occur.

Expansion: Product density (i.e., floating and sinking) can be controlled by extruder operation conditions and configuration.

Gelatinization: Extrusion cooking improves starch gelatinization.

Grinding: Ingredients can be ground in the extruder barrel during processing.

Homogenization: An extruder can homogenize by restructuring unattractive ingredients into more acceptable forms.

Mixing: A variety of screws are available which can cause the desired amount of mixing action in the extruder barrel.

Pasteurization and sterilization: Ingredients can be pasteurized or sterilized using extrusion technology for different applications.

Protein denaturation: Animal and plant protein can be denatured by extrusion cooking.

Shaping: An extruder can make any desired shape of product by changing a die at the end of the extruder barrel.

Shearing: A special configuration within the extruder barrel can create the desired shearing action for a particular product.

Texture alteration: The physical and chemical textures can be altered in the extrusion system.

Thermal cooking: The desired cooking effect can be achieved in the extruder.

Unitizing: Different ingredient lines can be combined into one product to produce special characteristics by using an extruder.

ADVANTAGES OF EXTRUSION

The principal advantages of extrusion technology compared to traditional food and feed processing methods based on Smith (1969) and Smith (1971) with modifications include the following:

Adaptability: The production of an ample variety of products is feasible by changing the minor ingredients and the operation conditions of the extruder. The extrusion process is remarkably adaptable in accommodating consumer demand for new products.

Product characteristics: A variety of shapes, textures, colors, and appearances can be produced, which is not easily feasible using other production methods.

Energy efficiency: Extruders operate with relatively low moisture while cooking food products, so therefore, less re-drying is required.

Low cost: Extrusion has a lower processing cost than other cooking and forming processes. Savings of raw material (19%), labor (14%), and capital investment (44%) when using the extrusion process have been reported by Darrington (1987). Extrusion processing also requires less space per unit of operation than traditional cooking systems.

New foods: Extrusion can modify animal and vegetable proteins, starches, and other food materials to produce a variety of new and unique snack food products.

High productivity and automated control: An extruder provides continuous high-throughput processing and can be fully automated.

High product quality: Since extrusion is a high-temperature/short-time (HT/ST) heating process, it minimizes degradation of food nutrients while it improves the digestibility of proteins (by denaturing) and starches (by gelatinizing). Extrusion cooking at high temperatures also destroys antinutritional compounds, i.e., trypsin inhibitors, and undesirable enzymes, such as lipases, lipoxidases, and microorganisms.

No effluent: This is a very important advantage for the food and feed industries, since new environmental regulations are stringent and costly. Extrusion produces little or no waste streams.

Process scale-up: Data obtained from the pilot plant can be used to scale up the extrusion system for production.

Use as a continuous reactor: Extruders are being used as continuous reactors in several countries for deactivation of aflatoxin in peanut meals and destruction of allergens and toxic compounds in castor seed meal and other oilseed crops.

DEVELOPMENT OF EXTRUDERS

Extrusion processes and extruders were developed simultaneously in various industries during the past two centuries (Janssen, 1978; Harper, 1981).

1797	Joseph Bramah, England, was the first to apply the extrusion principle by developing a hand-operated piston press to extrude seamless lead pipe. Similar equipment was later used for processing clay pipe, tile, soap, and pasta.
1869	Fellows and Bates, England, developed the first known continuous twin-screw extruder, originally used in sausage manufacture.
1873	Phoenix Gummiwerke A.G., Germany, developed the first known single-screw extruder, initially used for processing rubber.
Mid-1930s	Single-screw continuous pasta press was developed.
Late 1930s	Roberto Columbo and Carlo Pasquetti, Italy, adapted the twin-screw design for making plastics.
Late 1930s	General Mills, Inc., Minneapolis, MN, first used a single-screw extruder in the manufacture of ready-to-eat (RTE) cereals. Precooked, hot dough was shaped in an extruder before subsequent drying and flaking or puffing.
1939	Expanded corn curls or "collets" were first extruded. The product was not marketed until after World War II (1946) by the Adams Corporation, Beloit, WI.
1940	During the 1940s, a number of single-screw expellers, which squeeze the oil from oilseed, were developed and refined, replacing the use of much less efficient hydraulic presses previously employed for this purpose.
Late 1940s	Desires to improve appearance, palatability, and digestibility of animal feeds led to the development of the cooking extruder and the marketing of "Gaines Homogenized Meal," the first widely accepted modern dry dog food.
1950	Dry, expanded, extrusion-cooked pet foods quickly developed in the 1950s, largely replacing the biscuit baking processes which were used to manufacture them up to that time. The development of several new single-screw extruders expanded their application in the 1950s to commodity-type products such as dry pet foods, precooked cereal flours, and heat treated cereals and oilseeds to enhance their value as animal feed constituents.
Late 1950s	Pressurized preconditioners, which enable the precooking of ingredients above 212°F before entering the extruder screw, became available on Sprout-Waldron Company extruders (Muncy, PA). Precooking in a less costly section frees the closely-machined extruder for more shearing, working, and shaping functions.

1960s	Continuous cooking and forming of RTE cereals was developed as a one-step process on cooking extruders. Semimoist pet foods and precooked cereal food ingredients such as pregelatinized starches and cracker meals were marketed. Also, texturized soybean flour or concentrate products with a meat-like appearance were developed. These products are called "Texturized Plant Proteins" (TPP) and "Textured Soy Protein" (TSP) in the industry. The names "Textured Vegetable Protein©" and "TVP©" are copyrighted by the Archer Daniels Midland Company, Decatur, IL. "Dry" (autogenous) extruders were applied to trypsin inactivation of full-fat soybeans (InstaPro) and later to overseas "low cost extrusion" (LCE) needs.
Mid 1970s	Second generation (segmented screw and barrel cell) single- (Wenger, Sabetha, KS) and twin-screw extruders introduced Wenger and Creusot-Loire/Werner-Pfliderer, Germany.
Early 1990s	Conditioners, vented barrel, "third generation" deep flight reduced-autogenous heat extruders were introduced in feeds manufacture; annular gap extruders were retrofit with pellet mills.
1998	New generation extruders were patented by Wenger Manufacturing Co. (Sabetha, KS).

Simple, inexpensive extruders were initially developed in the United States in the 1960s for on-the-farm cooking of soybeans and cereal feeds in the 1960s. The main objective in processing soybeans was heat inhibition of the trypsin-inhibitor antigrowth factor, and other machines like the Gem Roaster and the Micronizer™ were also developed. The low-cost extruder designs were quickly adapted in the mid-1970s for use in nutrition intervention projects in many less-developed countries (LDCs) (Crowley, 1979). Numerous mechanical problems were experienced with early LCEs, but later models are more reliable and are widely used for processing different foods and crudely texturized foods in LDCs.

Twin-screw cooking extruders have been manufactured in Europe for over 35 years but did not attract significant interest in the United States until the early 1980s.

TERMINOLOGY

Each extruder manufacturer has its own special names/terms for their parts. Sometimes, the terminology is confusing and hard to understand. The following terms are most commonly used in the extrusion process:

- *Feedstock*—the material or mixture to be processed in an extruder
- *Preconditioner*—an assembly that adjusts moisture content and temperature of the ingredients and may partially or completely cook them before entering the extruder

- *Screw*—the member that conveys the product through the extruder
  - *flight*—the helical conveying surface of the screw which pushes the product forward
  - *pitch*—the angle of the flight, relative to the axis of the root
  - *root*—the solid or shaft part of the screw, around which the flight is wound
  - *worm*—a hollow-core, segmented screw element that slips over the shaft in a modular screw (Screws with various profiles and actions can be assembled by selecting appropriate worm sections.)
  - *shearlock, steamlock, shear ring, ring dam*—a ring-like device that locks together individual worm sections on a modular screw (Tight clearances induce shear and reduce blowback of steam to cooler sections of the barrel.)
  - *hollow-core screws*—solid screws or shaft may be drilled to circulate heating or cooling liquids, thus providing an extra heat transfer surface area
- *Shear*—a working, mixing action that homogenizes and heats the conveyed product
- *Interrupted- or cut-flight screw*—a screw with sections of flight missing (Usually, studs (bolts) are inserted through the extruder barrel wall into the empty flight section to induce shear in the product. Also, steam may be injected into the product through valves placed in the stud holes.)
- *Barrel*—a pipe-like retainer in which the extruder screw turns
- *Cooling/heating jacket*—a hollow sleeve around the barrel for circulating cooling water, or steam, or another heating medium like hot oil (In some locations, direct electrical heating of the barrel may be desirable.)
- *Vent*—an opening before the die plate in the extruder barrel which allows pressure and steam removal from the product
- *Barrel section*—a section of the barrel that is built in segments and may contain its own cooling/heating sleeve (The segments may contain grooving or spiraling, and often are the same length as the worm screws. In practice, different types of barrel sections are assembled to enhance the effects of the encompassed worm segment.)
- *Barrel liner*—a removable sleeve within the barrel (Usually a liner helps to resist the wear.)
- *L/D (Length-to-Diameter) ratio*—distance from the internal rear edge to the discharge end of the barrel, divided by the diameter of the bore (Food extruder L/D ratios range from 1:1 to 25:1.)

- *C.R. (Compression ratio)*—volume of the full flight of the screw at the feed opening, divided by the volume of the last full flight before discharge (Typical C.R. ranges are from 1:1 to 5:1.)
- *Die plate*—final assembly for shaping the product as it leaves the extruder (Die holes may be drilled directly into the plate or the plate may be machined to hold die inserts that have complex designs for shaping the product and may be made of hard-wearing material.)
- *Pellet*—discrete particle which is shaped and cut by an extruder, regardless of shape, sometimes referred to as a "collet"
- *Collet*—a word with many meanings—in oilseed extrusion, it is the coarse pieces made when extruding oilseeds to enhance their solvent extraction characteristics
- *Die land*—the constant-bore-length section of a die through which product passes (Longer lands give higher back pressures on the product and increase compression of the collet.)
- *Cutter*—assembly that cuts extrudate into pieces of desired length

Some terminology is illustrated in Figure 1.

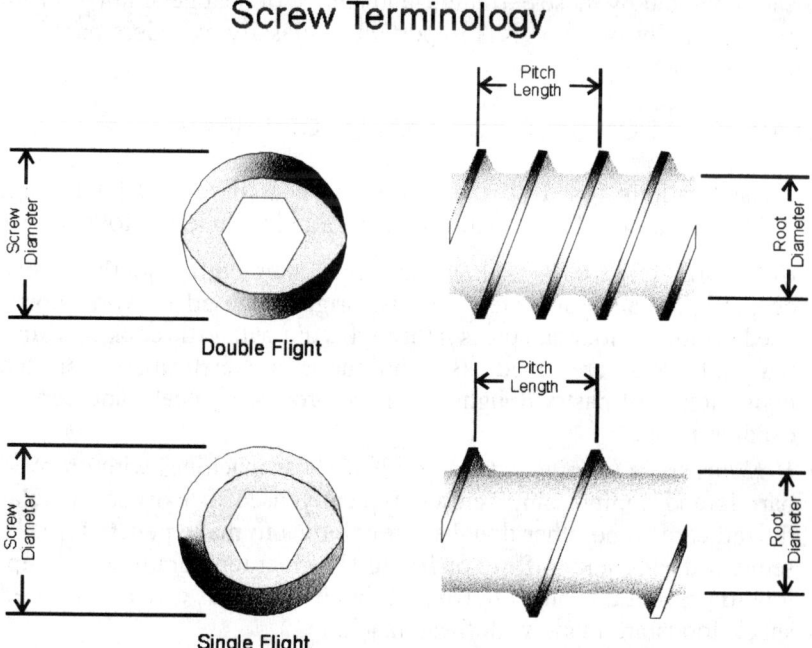

Figure 1 Screw terminology (courtesy of Wenger Manufacturing Co., Sabetha, KS).

TYPES OF EXTRUDERS

In today's food industry, the term "extruder" typically means a machine with Archimedean screw characteristics (i.e., a rotating flighted screw that fits tightly enough in a cylinder to convey a fluid) that continuously processes a product. Extruders may be designed to include various grinding, mixing, homogenizing, cooking, cooling, vacuumizing, shaping, cutting, and filling operations. Not all extruders are of the cooking-texturizing type. There are several different types of extruders available on the market. A few examples include dry extruders, interrupted-flight screw extruders, single-screw extruders, and twin-screw extruders.

Single-screw extruders are available in a number of sizes and shapes, and their screw, barrel, and die configurations can usually be varied to suit a particular product's specifications (Harper, 1978).

SINGLE SCREW CLASSIFICATION

A single-screw extruder can be classified based on several different characteristics, i.e., wet vs. dry, segmented vs. solid screw, extent of shear generated by these extruders, and source of heat generation. From a practical point of view, it is important to classify extruders based on shear and heat.

CLASSIFICATION BASED ON EXTENT OF SHEAR

Classifications based on the extent of shear described by Farrell, (1971) and Harper (1981), with modifications, include the following:

- *Cold forming extruders*—low-shear machines with smooth barrels, deep flights, and low screw speeds, originally used to work moistened semolina flour and press it through a die with little cooking (Similar extruders are used as continuous mixer-formers for the manufacture of pastry doughs, cookies, processed meats, and certain candies.)
- *High-pressure forming extruders*—low-shear machines with grooved barrels and compressing screws, typically used to extrude pregelatinized cereal and other doughs through dies to make pellets for subsequent drying and puffing or frying (Product temperatures are kept low to prevent unwanted puffing at the die. Various cereals and fried snack foods are made with these machines.)
- *Low-shear cooking extruders*—moderate-shear machines with high compression screws and grooved barrels to enhance mixing [Heat can

be applied to the barrel or screw to "cook" the product (pasteurize bacteria, inactivate enzymes, denature proteins, gelatinize starch), but puffing at the die is avoided. Soft-moist foods and meat-like snacks such as simulated jerky can be made with these machines. The ingredients are often premixed to a dough-like consistency using other equipment.]

- *Collet extruders*—high-shear machines with grooved barrels and screws with multiple shallow flights that have been used for making puffed snacks from defatted corn grits [The temperature of relatively dry (12% moisture) ingredients is raised rapidly to over 175°C, and the starch is dextrinized and partially gelatinized. The resulting mass loses moisture and puffs immediately upon exit through a die to form a crisp, expanded curl or collet. This type of machine initially was characterized by an extremely short screw (length: diameter = 3:1), but longer L/D (1:10) machines that rely heavily on friction-induced heat to produce collets have been developed. An imported "collet-type" short L/D extruder is offered domestically for processing animal feeds.]

- *High shear cooking extruders*—high-shear machines, with screws for changing flight depth and/or screw pitch, that have the ability to achieve high compression ratios, high temperatures, and various degrees of puffing [Long barrel (length diameter = 15–25:1) extruders adapted from the plastics industry were used initially, but many design modifications have been introduced for processing foods. A large variety of screw and internal barrel designs and heating and cooling options exist. Some machines are equipped with conditioning chambers to premoisten and preheat the feedstock material. Smith (1976) and others (Linko et al., 1981) have termed extrusion cookers designed to minimize the time that materials are held at maximum temperature as "high-temperature/short-time "(HT/ST) devices. Since heat and pressure cause the ingredients to flow during processing, this type of extrusion cooking has also been called "thermoplastic extrusion" (Last, 1979).

High-shear cooking extruders have been used for one-step preparation of RTE cereals (Kent, 1975), snack foods (Lachman, 1969; Matz, 1976; Inglett, 1975; Gutcho, 1973; Duffy, 1981; Stauffer, 1983; Matson, 1982), candy (Groves, 1982), crispbreads (Anderson et al., 1981), dry expanded pet foods (Horn and Bronikowski, 1979), precooked food ingredients such as pregelatinized corn and sorghum grits (Anderson et al., 1969a; Anderson et al., 1969b), pregelatinized corn flours (Smith et al., 1979) and alkali-treated (masa) corn flours for ethnic foods (Bazua et al., 1979; Bedolla and Rooney, 1982), dried soup mixes, instant beverage powders, croutons and breadings, crackers and wafers (Hauck, 1980), enzyme deactivation of full-fat soy flour (Mustakas and Griffin, 1964; Mustakas et al., 1971; Smith, 1969), imitation nuts (Moore and Rice, 1981), texturization of soy proteins (Smith, 1976; Kinsella, 1978), famine relief feeding (de Muelenaere and Buzzard, 1969; Crowley, 1979), and deactivation of enzymes in cereals and oilseeds (Sayre et al., 1982). The use of cooking extruders in the manufacture of breakfast cereals is described in the book edited by Fast and Caldwell (1990).

CLASSIFICATION BASED ON HEAT GENERATION

Classifications have also been based on how the feedstock is heated in a single-screw extruder during processing (Rossen and Miller, 1973).

Adiabatic (autogenous) extruders develop essentially all heat by friction (viscous dissipation of mechanical energy input), and little if any heat is removed through the barrel. Examples include "dry extruders," "collet extruders," and "low-cost" extruders used in LDC programs. Some extruders need to be heated by supplementary sources initially, but then will operate autogenously. Adiabatic extruders operate at low moisture levels (8–14%).

For more detailed information about these types of extruders, see Chapter 3, "Dry Extruders."

Isothermal extruders operate at an essentially constant product temperature throughout the entire length of the barrel and are used mainly for forming. Water-cooled jackets are sometimes used.

Polytropic extruders have provisions for alternately adding or removing heat as required by the specific process. Examples include most cooking extruders with external heating and cooling sections, which generate heat by friction.

The single-screw extruder can be categorized on the basis of its design. There are several different designs available on the market for the single-screw extruder. The following are three different designs, which are most commonly used in the food/feed industry. All of these types of extruders offer advantages in regard to their design.

SOLID SINGLE-SCREW EXTRUDERS

The classical drawing of a solid-screw extruder is shown in Figure 2, which can be used to explain the principles of how an extruder works. Note that the area of product volume is decreased from the feed to the discharge end of the screw by the thickening of the screw root, which results in shallower, fixed pitch flights. If not already preconditioned, the dry ingredients are moistened and mixed early in the first feed sec-

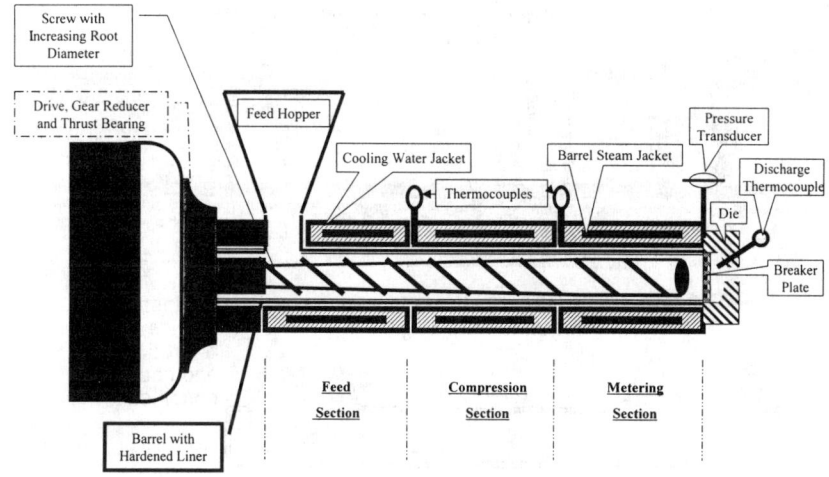

Figure 2 Single solid screw extruder—basic components.

tion, and then they are compressed in the transition section, cooked in the metering section, discharged through a shaping die, and cut to desired lengths by a suitable rotating knife. Compression in the transition zone can be as high as 5:1. The extruder in Figure 2, is also equipped with several jackets that allow heating and cooling of the barrel. For example, in operation, the section next to the feed end might be cooled to maintain product viscosity and prevent blowback of steam from the cooking section. The barrel next to the die end might be kept hot if an expanded product is desired, or cold (to reduce product temperature below the boiling point of water) if expansion is not wanted. Shear occurs as the compressed product is wiped against the wall of the extruder barrel and fed forward against the back pressure created by the die plate.

INTERRUPTED-FLIGHT EXTRUDER-EXPANDER

Figure 3 shows an interrupted-flight screw extruder-expander. The flight is not continuous, but has a section missing into which a shear bolt is inserted through the jacket wall. Heat is induced by friction as the product is conveyed past the shear bolts by the screw. These machines may or may not be jacketed, but are often equipped for direct injection of steam. They are also referred to as "cut flight" extruders or as "expanders" after the original Anderson Grain Expander was introduced in the mid-1950s. Back pressure in the machine is induced by

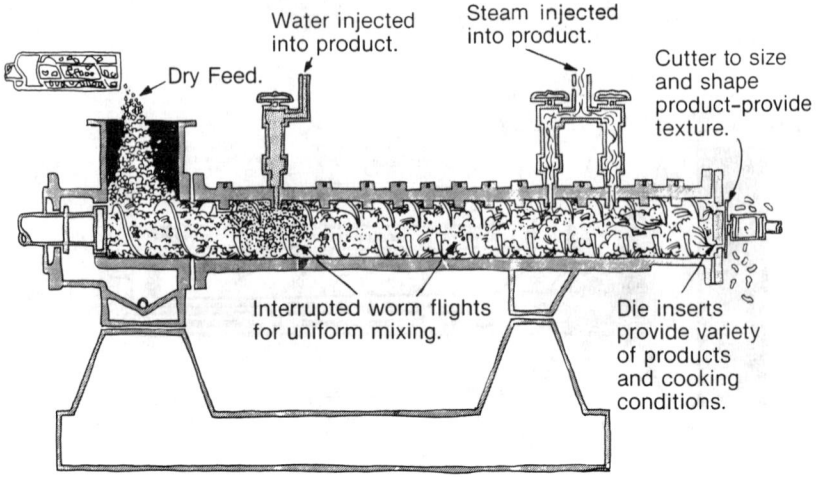

Figure 3 Cross section of interrupted-flight expander (courtesy of Anderson International Corp., Cleveland, OH).

restriction at the die. A raw product plug is also created at the feed end to enclose a "reactor cell."

Although the barrel is one piece, the flight is manufactured in small sections that slip over a keyed, round shaft. The worn flight sections can be easily replaced or switched toward the back of the screw. Additionally, the flights near the discharge end of the machine can be surfaced with abrasion-resistant alloys to extend their useable life. The segmented flight principle was borrowed from screw presses invented by V. D. Anderson in the 1890s. The interrupted-flight design avoids spinning of the product with the screw and enables the use of a smooth-wall barrel. For more details, see Chapter 4, "Interrupted-Flight-Exturders-Expanders."

SINGLE, SEGMENTED-SCREW EXTRUDERS

A cross section of a single-screw extruder with a segmented worm screw and barrel is shown in Figure 4 . Note that the root of the screw is constant in diameter, and that compression results by changing the pitch of the worm flights. In this case, the full-pitch screw segment forces the product onto a worm with twin-spiral one-half pitch flights to achieve compression. The degree of shear can be modified in two ways: by using shear locks of increasing diameters, requiring increasing pressures to force the product into the adjacent worm section; and by selecting between a grooved- and a spiral-walled barrel. More mix-

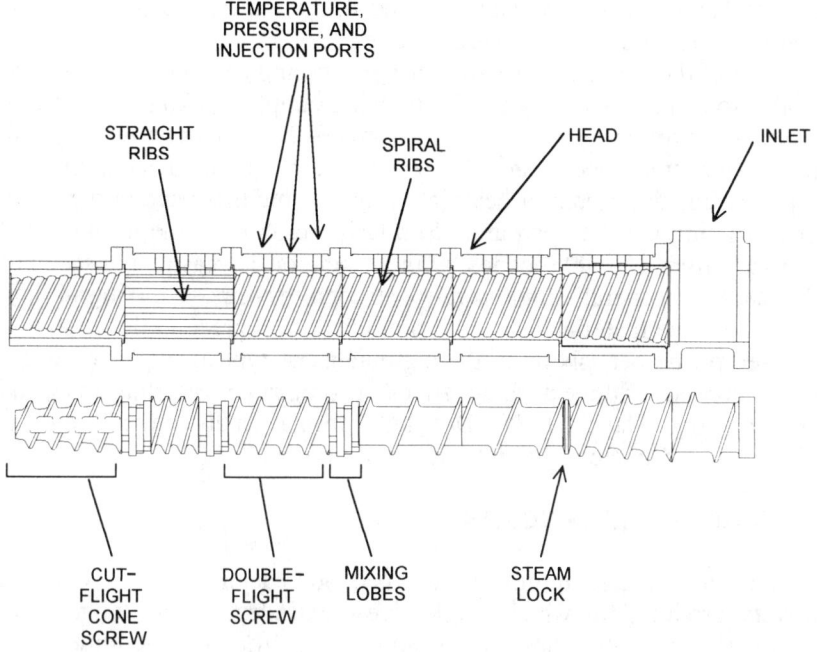

Figure 4 Cross section of a segmented single-screw extruder (courtesy of Wenger Manufacturing Co., Sabetha, KS).

ing and shearing occurs with the grooved barrel because of greater slippage between the screw flight and barrel walls. Straight and spiral walls are also shown in Figure 4. For more details, see Chapter 2, "Single-Screw Extruders."

Today's major large capacity single-screw extruders are primarily of the segmented screw and barrel design. The significance of the segmented screw and barrel design is that each section between two shearlocks becomes a reactor cell, instead of one reactor cell as in the solid-screw and interrupted-flight extruders described earlier. The number of successive cells that can be put on a screw is only limited by its length. Five, seven, and nine cells are common. A different operation can be conducted in each cell, including compression, jacket or steam injection heating, shearing, venting of steam and product cooling, recompression, addition of heat-sensitive ingredients, and jacket cooling.

Solid-screw extruders are still built for special low-pressure cooking and forming applications. But, segmented screw worms and barrel sections offer considerable versatility in customizing screw-barrel design and replacing worn parts and are used in the majority of single-screw and twin-screw extruders that are built.

Most of the single-screw extruder's processing conditions can be controlled to achieve a variety of effects. For example, cooking temperature within the extruder barrel can range from 80–200°C by configuring with high shear screws and shearlocks, injecting direct steam, heating the barrel by circulating steam or heating oil, increasing the speed of the shaft, or restricting the die open area. Similarly, residence time in the barrel can vary from 15–300 seconds by increasing or decreasing the speed of the shaft. In general, single-screw extruders have poor mixing ability. Therefore, the material should be premixed, or a preconditioner should be used for proper mixing of the ingredients. A typical single-screw extruder consists of three different zones: feeding zone, kneading zone, and cooking zone (Hauck, 1985). Detailed information about these zones is discussed in Chapter 2 "Single-Screw Extruders."

TWIN-SCREW EXTRUDERS

In recent years, requirements have been increasing for new and higher quality products for which single-screw extruders are no longer adequate. However, for these more demanding processing requirements, twin-screw technology must be used. Twin-screw extruders include a variety of machines with widely different processing and mechanical characteristics and capabilities. Most of the improvements that have evolved in the development of extruders have been incorporated into the modern twin-screw extruders.

CLASSIFICATION OF TWIN-SCREW EXTRUDERS

Twin-screw extruders can be classified on the basis of direction of screw rotation in the following two categories.

(1) Counterrotating twin-screw extruders
(2) Corotating twin screw extruders

These categories can be further subdivided on the basis of position of the screws in relation to one another into the following: intermeshing and nonintermeshing.

COUNTERROTATING TWIN-SCREW EXTRUDERS

These types of extruders are not widely used in the food industry although they are excellent conveyors. They are good in processing relatively nonviscous materials requiring low speeds and long residence times. Examples are gum, jelly, and licorice confections (Elsner and Wiedmann, 1985).

COROTATING TWIN-SCREW EXTRUDERS

These types of extruders are most commonly used in the food and snack food industry. Corotating twin-screw extruders have played a major role in broadening the variety of products that can be made using extrusion technology. These types of extruders provide high degrees of heat transfer but not forced conveyance (Elsner and Wiedmann, 1985). Advantages of this type of system include its pumping efficiency, good control over residence time distribution, self-cleaning mechanism, and uniformity of processing (Schuler, 1986).

As shown in Figure 5, the screw either rotates in opposing directions (counterrotating) or in the same direction (corotating).

Extruders can have the following screw positions:

(1) *Intermeshing screw*—In this extruder, the flight of one screw engages or penetrates the channels of the other screw. Positive pumping action, efficient mixing, and self-cleaning characteristics are offered. These features distinguish them from single-screw and nonintermeshing screw extruders.

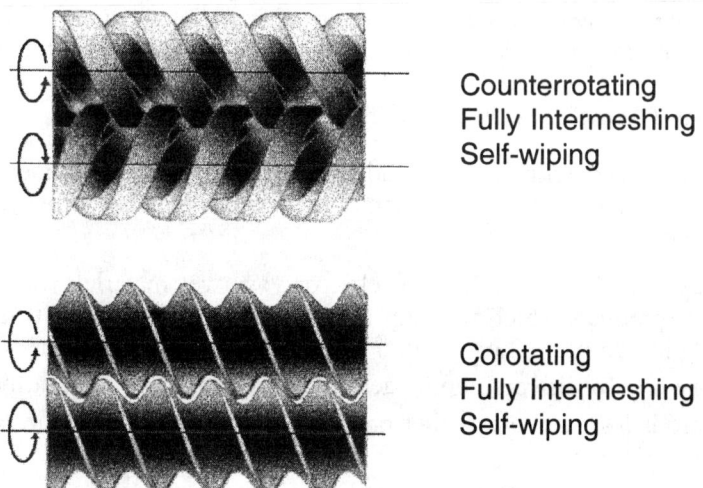

Counterrotating
Fully Intermeshing
Self-wiping

Corotating
Fully Intermeshing
Self-wiping

Figure 5 Counter- and corotating fully intermeshing self-wiping twin-screw (courtesy of Wenger Manufacturing Co., Sabetha, KS).

(2) *Nonintermeshing Screw*—In this extruder, the screws do not engage each other's threads, allowing one screw to turn without interfering with the other. Clark (1978) described the nonintermeshing screw extruders as two single-screw extruders sitting side by side with only a small portion of the barrel in common. Like the single-screw extruder, these extruders depend on friction for extrusion. These screws are not designed for pumping or mixing purposes. Nonintermeshing screw extruders function like single-screw extruders, but they have a higher capacity (Dziezak, 1989).

According to Miller (1990), four types of twin-screw extruders are possible.

- nonintermeshed, corotating
- nonintermeshed, counterrotating
- intermeshed, corotating
- intermeshed, counterrotating

Extruders have been built with all of these variations. It has been noted that nonintermeshed twin-screw extruders may act as two separate screws laying side by side, with uneven filling and discharge from each screw. Self-wiping versions of the corotating intermeshed twin-screw extruders are very popular domestically. However, interest is growing in processing materials that require high pumping pressures in intermeshed counterrotating twin-screw extruders. Compared to intermeshing counterrotating screws, intermeshing corotating screws transport four to five times more volume of material in open, V-shaped chambers. For more details see Chapter 5, "Twin-Screw Extruders."

ADVANTAGES OF TWIN-SCREW EXTRUDERS

Intermeshed, twin-screw extruders typically cost 50–150% more than single-screw extruders of the same throughput, but they offer several advantages.

- They handle viscous, oily, sticky, or very wet materials and some other products which will slip in a single-screw extruder. (It is possible to add up to 25% fat in a twin-screw extruder.)
- They have positive pumping action and reduced pulsation at the die.
- There is less wear in smaller parts of the machine than in the single-screw extruder.
- They feature a nonpulsating feed.
- A wide range of particle size (from fine powder to grains) may be used, whereas single-screw is limited to a specific range of particle size.

- Cleanup is very easy because of the self-wiping characteristics.
- The barrel head can be divided into two different steams.
- They provide for easier process scale-up from pilot plant to large-scale production.
- Their process is more forgiving to inexperienced operators.

A cross section of a corotating, self-wiping twin-screw extruder screw is shown in Figure 6. The basic elements of feeding, kneading, and cooking zones are still there, as are the shearlocks.

Books by Frame (1994), Mercier et al. (1989), O'Connor (1987), Jowitt (1984), Harper (1981), and Janssen (1978), and articles by Fichtali and van de Voort (1989), Hauck and Huber (1989), Midden (1981),

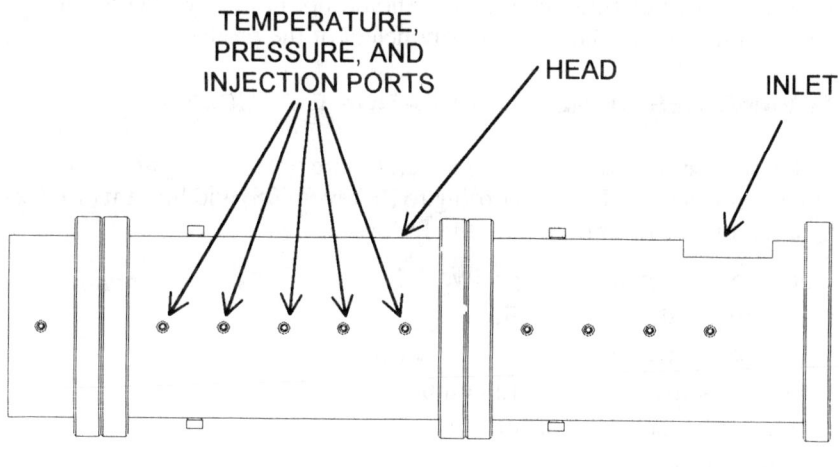

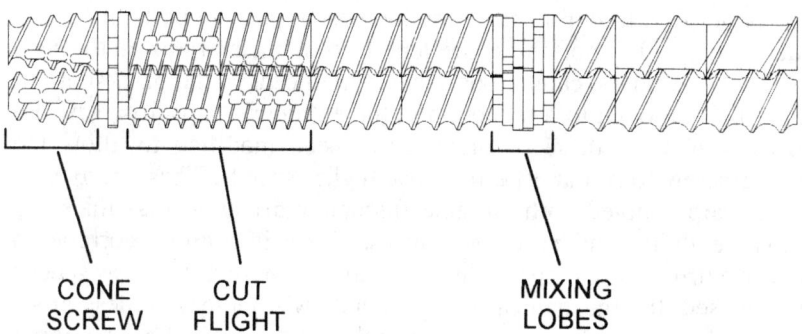

Figure 6 Cross-sectional view of twin-screw extruder (courtesy of Wenger Manufacturing Co., Sabetha, KS).

Padmanabhan and Bhattacharya (1989), Dziezak (1989), Miller, (1985a; 1985b), Levine (1988), Mulvaney and Hsieh (1988), Lusas and Rhee (1987), Harper (1986), Schuler (1986), Lazarus and Renz (1985), Miller (1985), Straka (1985), Woollen (1985), Faubion et al. (1982), Tribelhorn and Harper (1980), and Rossen and Miller (1973), describe design and operating characteristics of single- and twin-screw extruders.

NEW GENERATION EXTRUDERS

New generation single-screw extruders were patented by Wenger Manufacturing Co., Sabetha, Kansas, in 1998. These systems are designed to operate at high shaft speeds and small length-to-diameter ratios. In order to be successful in today's competitive market, any new system must meet or exceed exiting benefits and should not have any adverse effect on the nutrition or other quality parameters of the product.

ADVANTAGES OF NEW GENERATION EXTRUDERS

These types of extruders offer several more potential advantages than other types of extruders. According to Rokey (1998) and Strahm (1999), benefits of this design include the following:

- 30 to 50% increase in capacity
- 5 to 20% reduction in bulk density
- \> 25% reduction in energy consumed
- reduced sensitivity to worn components
- improved processing of high carbohydrate diets
- reduced processing and capital cost

Since these extruders have shorter L/D ratios, this means there are fewer stationary and rotating wear parts. The fixed and rotating components of the extruder's barrel are the most expensive parts of the extruder system. By shortening the length-to-diameter ratio of the new generation extruders, capital costs can be reduced. In the past, shortening the L/D ratio led to a reduction in the extruder's capacity. New derive assemblies in these extruders have been modified to allow the extruder screw to run at 30% to 100% higher speeds. These derive assemblies are coupled with variable frequency drives to maximize extrusion flexibility and product controls. There is a direct correlation between extruder screw speeds and mechanical energy. As screw speeds are increased, the mechanical energy inputs will increase unless product throughputs are also increased to maintain the desired energy input per unit of throughput. Simply increasing the mechanical energy per unit of throughput will decrease the bulk density of extruded products. New generation extruders also demonstrate reduced sensitivity to worn components and enhanced product separation at the die when carbohydrate diets are utilized. Increased screw speed reduced bulk density in high carbohydrate (rice) diets without the clumping of product at the die that usually results from these starchy diets. In new generation extruders, AC variable frequency drives can enhance the flexibility to match the flexibility of a twin-screw extruder in many processes. Since higher capacities are reached on the new generation systems at approximately the same level of mechanical energy input, higher power extruder derive are required.

Greater output at less cost continues to be the major driving force behind most technological advances. The new generation extruders are no exception to this trend. The operating costs for this system can be broken down into the following categories: raw material, capital (depreciation and interest), maintenance, labor, and utilities. In Table 1, these costs (excluding raw materials) are itemized on a per ton basis for a modern single-screw system (present technology); a new generation single-screw system with variable speed screw, and a modern twin-screw system. Each extrusion system includes extruder, dryer, and manual control systems (Table 1).

The new generation extruders decrease the total operating costs primarily due to a reduction in capital costs as well as increased energy efficiency. With about a quarter per ton additional investment, the new generation extruders' capabilities can be expanded to match those of a twin-screw extruder which has an inherently higher operating cost (Table 2).

TABLE 1. **Summary of Operating Costs for Extrusion Systems.**

Cost Category	Units	Present Single-Screw Technology	Single-Screw New Generation	Single-Screw New Generation with/VS Drive	Twin-Screw
Production Rate	mt/hr	11	11	11	8
Capital Cost	$/mt	2.07	1.68	1.88	3.25
Maintenance	$/mt	0.54	0.52	0.52	2.01
Labor	$/mt	1.82	1.82	1.82	2.50
Utilities	$/mt	5.90	5.59	5.59	6.89
Miscellaneous	$/mt	0.18	0.15	0.16	0.28
Total Cost		10.51	9.75	9.97	14.93

Courtesy of Wenger Manufacturing Co., Sabetha, KS.

TABLE 2. Capital Cost Comparison (11 ton/hr System).

	Existing Technology	New Generation Extruders
Investment Cost ($/mt.)	0.61	0.56
Depreciation ($/mt.)	1.92	1.75
Total Capital Cost ($/mt.)	2.54	2.31
Utilities ($/mt.)	7.04	6.36
Maintenance ($/mt.)	0.90	0.67

Courtesy of Wenger Manufacturing Co., Sabetha, KS.

Although the basic principles are still applicable, several earlier extruder classification systems have been made obsolete by new designs. With the general consolidation of food and feed manufacturing operations into larger, centralized, highly automated installations during the last two decades, many earlier small throughput machines have been set aside and are now mainly of historical interest. Whereas extruders were once designed for very specialized purposes with solid-screw, barrel, and die designs, the current practice is to build a basic drive assembly that is then outfitted with combinations of modular preconditioners, screw worms, barrel sections, dies, and cutters to obtain the desired shearing, heating/cooling, and product shaping effects desired. In this evolution, the early "collet-type" extruders for making puffed snack foods have been mainly replaced by short L/D cooking-type extruders often operated adiabatically, and many cold-forming pasta presses are being replaced by twin-screw extruders that cook and shape in one step.

SUGGESTED READINGS

Chang, Y. K. and S. S. Wang. 1999. *Advances in Extrusion Technology.* Aquaculture/Animal Feeds and Foods. Lancaster, Pennsylvania: Technomic Publishing Co.

Fast, R. B. and E. Caldwell. 1993. *Breakfast Cereals and How They are Made.* American Association of Cereal Chemists, St. Paul, MN.

Frame, N. D. 1994. *The Technology of Extrusion Cooking.* New York: Blackie Academic & Professional.

Harper, J. M. 1981. *Extrusion of Foods.* Vol. 1 and 2. Boca Raton, Florida: CRC Press, Inc.

Hayakawa, I. 1992. *Food Processing by Ultra High Pressure Twin Screw Extrusion.* Lancaster, Pennsylvania: Technomic Publishing Co.

Janssen, L. P. B. M. 1978. *Twin Screw Extrusion.* New York: Elsevier Applied Science.

Jowitt, R. 1984. *Extrusion Cooking Technology.* New York: Elsevier Applied Science.

Kokini, J. L., C. Ho, and M. V. Karwe. 1992. *Food Extrusion Science and Technology.* New York: Marcel Dekker, Inc.

Mercier, C. and C. Cantarelli. 1986. *Pasta and Extrusion Cooked Foods.* New York: Elsevier Applied Science.

Mercier, C., P. Linko, and J. M. Harper. 1989. *Extrusion Cooking.* American Association of Cereal Chemists, St. Paul, MN.

O'Connor, C. 1987. *Extrusion Technology For the Food Industry.* New York: Elsevier Applied Science.

Pet Food Industry. 1999. *Focus on Extrusion* (Proceedings). Mt. Morris, IL: Watt Publishing Co.

Wilson, D. and R. E. Tribelhorn. 1979. *Low-cost Extrusion Cookers.* Workshop Proceedings. United States Department of Agriculture. Office of International Cooperation and Development, Washington D. C.

Woodroofe, J. M. 1993. *Dry Extrusion Manual.* Rural Pacific Pty., Ltd., Australia.

Zeuthen, P., J. C. Cheftel, C. Eriksson, M. Jul, H. Leniger, P. Linko, G. Varela, and G. Vos. 1984. *Thermal Processing and Quality of Foods.* New York: Elsevier Applied Science.

REFERENCES

Anderson, R. A., H. F. Conway, V. F. Pfeifer, and E. W. Griffin, Jr. 1969a. "Gelatinization of corn grits by roll- and extrusion-cooking." Cereal Science Today. 14: 4–7, 11–12.

Anderson, R. A., H. F. Conway, V. F. Pfeifer, and E. W. Griffin, Jr. 1969b. "Roll and extrusion-cooking of grain sorghum grits." Cereal Science Today. 14: 372–375, 381.

Anderson, Y., B. Hedlund, L. Jonsson, and S. Svensson. 1981. "Extrusion cooking of a high-fiber cereal product with crispbread character." Cereal Chem. 58: 370–374.

Bazua, C. D., R. Guerra, and H. Sterner. 1979. "Extruded corn flour as an alternative to lime-heated corn flour for tortilla preparation." J. Food Sci. 44: 940–941.

Bedolla, S. and L. W. Rooney. 1982. "Cooking maize for masa production." Cereal Fds. World. 27: 218–222.

Clark, J. P. 1978. "Texturization by extrusion." J. Texture Studies. 9: 109.

Crowley, P. R. 1979. "Transferring LEC technology to developing countries: from concept to application and beyond." In: *Low-Cost Extrusion Cookers, Second International Workshop Proceedings* (Tanzania). D. E. Wilson and R. E. Tribelhorn, eds., Dept. Agr. and Chem. Engr., Colorado State University, Ft. Collins, CO, pp. 11–14.

Darrington, H. 1987. "A long-running cereal." Food Manuf. 3: 47–48.

de Muelenaere, H. J. H. and J. L. Buzzard. 1969. "Cooker extruders in service of world feeding." Fd. TechnoL. 23: 345–351.

Duffy, J. I. 1981. *Snack Food Technology: Recent Developments,* Park Ridge, NJ: Noyes Data Corporation.

Dziezak, J. D. 1989. "Single and twin-screw extruders in food processing." Fd. TechnoL. 43(4): 163–174

Elsner, G. and W. Wiedmann. 1985. "Cooker extruder for the production of gums and jelly articles." Impulse Food Suppl. Nov., p.2.

Farrell, D. 1971. "Extrusion equipment—types, functions and applications." Symposium on Extrusion: Process and Product Development, American Association of Cereal Chemists, St. Paul, MN.

Fast, R. B. and E. F. Caldwell. 1990. "Unit operations and equipment. IV. Extrusion and extender." In: *Breakfast Cereals and How They Are Made,* American Association of Cereal Chemists, St. Paul, MN, pp. 135–193.

Faubion, J. M., R. C. Hoseney, and P. A. Seib. 1982. "Functionality of grain components in extrusion." Cereal Fds. World. 27: 212–216.

Fichtali, J. and F. R. van de Voort.1989. "Fundamental and practical aspects of twin-screw extrusion." Cereal Fds. World. 34: 921–929.

Frame, N. D. 1994. *The Technology of Extrusion Cooking.* Glasgow: Pub. Blackie Academic & Professional. Chapman & Hall.

Groves, R. 1982. "Applications for cereal in candy manufacturing." Cereal Fds. World. 27: 589–591.

Gutcho, M. 1973. *Prepared Snack Foods,* Park Ridge, NJ: Noyes Data Corporation.

Harper, J. M. 1978. "Extrusion processing of food." Fd TechnoL. 32(7): 67–72.

Harper, J. M. 1981. *Extrusion of Foods.* Vols. I and II. Boca Raton, FL: CRC Press, Inc.

Harper, J. M. 1986. "Extrusion texturization of foods." Fd. TechnoL. 40(3): 70–76.

Hauck, B. W. 1980. "Marketing opportunities for extrusion cooked products." Cereal Fds. World. 25: 594–595.

Hauck, B. W. 1985. "Comparison of single and twin screw cooking extruders." Impluse Food Suppl. Feb., p. 6

Hauck, B. W. and G. R. Huber. 1989. "Single screw vs twin screw extrusion." Cereal Fds. World. 34: 930, 932–934, 936–939.

Horn, R. E. and J. C. Bronikowski. 1979. "Economics of extrusion processing." Cereal Fds. World. 24: 140–141.

Inglett, G. E. 1975. *Fabricated Foods.* Westport, CT: AVI Publishing Co.

Janssen, L. P. B. M. 1978. *Twin Screw Extrusion.* New York: Elsevier.

Jowitt, R. 1984. *Extrusion Cooking Technology.* New York: Elsevier.

Kent, N. L. 1975. *Technology of Cereals.* New York: Pergamon Press.

Kinsella, J. E. 1978. "Texturized protein: fabrication, flavoring, and nutrition." CRC Crit. Rev. Fd. ScL Nutr. 10(2): 141–206.

Lachman, A. 1969. *Snacks and Fried Products.* Park Ridge, NJ: Noyes Development Corporation.

Last, J. 1979. "Thermoplastic extrusion trials of some oilseed, legume and cereal proteins." CSIRO FL Res. Qtr. 39: 25–29

Lazarus, C. R. and K. H. Renz.1985. "The influence of cereal flours on the taste perception of extrusion-stable flavors." Cereal Fds. World. 30: 319–320 .

Levine, L. 1988. "Understanding extruder performance." Cereal Fds. World. 33: 963–970.

Linko, P., P. Colonna, and C. Mercier. 1981. "High temperature short-time extrusion." In: *Advances in Cereal Science and Technology,* Vol. IV, Y. Pomeranz, ed., American Association of Cereal Chemists, St. Paul, MN, pp. 145–235.

Lusas, E. W. and K. C. Rhee. 1987. "Extrusion processing as applied to snack foods and breakfast cereals." In: *Cereal and Legumes in the Food Supply.* J. Dupont and E.M. Osman, eds., Iowa State University Press, Ames, IA, pp. 201–218.

Matson, K. 1982. "What goes on in the extruder barrel." Cereal Foods World. 27:207–210.

Matz, S. A. 1976. *Snack Food Technology.* Westport, CT: AVI Publishing Co.

Mercier, C., P. Linko, and J. M. Harper. 1989. *Extrusion Cooking.* American Association of Cereal Chemists, Inc., St. Paul, MN.

Midden, T. M. 1981. "Twin screw extrusion of corn flakes." Cereal Fds. World. 34: 941–943.

Miller, R. C. 1985a. "Extrusion cooking of pet foods." Cereal Fds. World. 30: 323–327.

Miller, R. C. 1985b. "Low moisture extrusion: effects of cooking moisture on product characteristics." I. Food ScL. 50: 249–253.

References

Miller, R. C.,1990. "Unit operations and equipment. IV. Extrusion and Extruders." In: *Breakfast Cereals and How They Are Made*. R. B. Fast and E. F. Caldwell, eds., American Association of Cereal Chemists, St. Paul, MN, pp. 135–193.

Moore, K. and J. Rice. 1981. "American peanut crop cut in half, processors search for acceptable substitutes." Food Processing. 42(1): 64–67.

Mulvaney, S. and F-H. Hsieh. 1988. "Process control for extrusion processing." Cereal Fds. World. 33: 971–976.

Mustakas, G. C., W. Albrecht, G. N. Bookwalter, V. E. Sohns, V. E. and E. L. Griffin, Jr. 1971. "New process for low-cost, high protein beverage base." Fd. TechnoL. (25): 534–540.

Mustakas, G. C. and E. L. Griffin, Jr. 1964. "Production and nutritional evaluation of extrusion-cooked full-fat soybean flour." I. Am. Oil Chem. Soc. 41: 607–614.

O Connor, C. 1987. *Extrusion Technology for the Food Industry*. New York: Elsevier Applied Science.

Padmanabhan, M. and M. Bhattacharya. 1989. "Extrudate expansion during extrusion cooking of foods." Cereal Fds. World. 34: 945–949.

Rokey, G. 1998. "New processing technologies." Paper presented at Petfood Forum 98. Chicago, IL, Mar. 30–Apr. 1.

Rossen, J. L. and R. C. Miller. 1973. "Food extrusion." Food Technol. 27(8): 46–53.

Sayre, R. N., R. M. Saunders, R. V. Enochian, W. G. Schultz, and E. C. Beagle. 1982. "Review of rice bran stabilization systems with emphasis on extrusion cooking." Cereal Fds. World. 27: 317–322.

Schuler, E. W. 1986. "Twin-screw extrusion cooking for food processing." Cereal Fds. World. 31: 413–416.

Smith, O. B. 1969. "History and status of specific protein-rich foods: extrusion-processed cereal foods." In: *Protein-Enriched Cereal Foods for World Needs*. M. Milner, ed. American Association of Cereal Chemists, St. Paul, MN, pp. 140–153.

Smith, O. B. 1971. "Why use extrusion." Symposium on Extrusion: Process and Product Development. American Association of Cereal Chemists. St. Paul, MN.

Smith, O. B. 1976. *Extrusion cooking. New Protein Foods*. N. M. Altschul, ed., New York: Academic Press, pp. 86–121.

Smith, O. and T. S. Debuckle, N. M. Desandoval, and G. E. Gonzales.1979. "Production of precooked corn flours for arepa making using an extrusion cooker." J. Food Sci. 44: 816–819.

Stauffer, C. E. 1983. "Corn-based snacks." Cereal Fds. World. 28: 301–302.

Strahm, B. 1999. "New generation extruders." In: *Proceedings of Focus on Extrusion by Pet Food Industry*. Mt. Morris, IL: Watt Publishing Co., p. 24–28.

Straka, R. 1985. "Twin- and single-screw extruders for the cereal and snack industry." Cereal Fds. World. 30: 329–332.

Tribelhorn, R. E. and J. M. Harper. 1980. "Extruder-cooker equipment." Cereal Fds. World. 25: 154–156.

Woollen, A.1985. "Higher productivity in crisp bread." Cereal Fds. World. 30: 333–334.

CHAPTER 2

Single-Screw Extruders

GALEN J. ROKEY

EXTRUSION has been simply defined as the process of forcing a material through a specifically designed opening. Extrusion as a process has been known since the late eighteenth century. Joseph Bramah in 1797 in England built a hand-operated piston press for lead pipes. The development of continuously operated extruders for rubber took place in the middle of the nineteenth century in England and Germany. Bewly and Brooman obtained a patent in 1845 for a hand-operated extruder which was converted in 1855 to a mechanically driven extruder. In 1879, Shaw in England and Royle in the U.S. in 1880, made a single-screw extruder for rubber.

The first food extruders were based on the use of piston and ram. Single-screw type extruders used for chopping or mincing soft food by forcing them through die plates may have been the first screw extruders used in the food industry. In Italy, single-screw extruders were used in the mid-1930s for pasta products. The principle of these extruders remains the same with recent developments focused on increased capacity and improved control. They employ low shear, deep-flight screws and operate at low screw speeds. Around the same time in the U.S., similar extruders were used in the breakfast cereal industry to form precooked cereal dough (Yacu, 1999).

Commercial extrusion processing of food and feeds has been practiced for nearly sixty years. The first commercial application of screw extruders in the food industry was the production of pasta using a single-screw device. This product was not fully cooked. Moist dough was compressed by the slow-turning screw and shaped by the orifice through which the dough was expelled. The screw extruder was first used as a continuous cooking device in the late 1930s. In the mid-1940s, the first commercial application of the extrusion process was the production of expanded cornmeal snacks.

In the early 1950s, extrusion cookers were first applied to the production of dry, expanded pet foods. Today, the extrusion cooker has become the primary continuous cooking apparatus in the commercial production of pet foods. Now, pet foods account for the largest annual tonnage of extrusion cooked product in the U.S. and undoubtedly the entire world.

The extrusion process was made continuous by substituting the piston in the cylinder of the original design with a helical screw. In this screw extrusion apparatus, material is continuously metered into an inlet hopper and then transported forward by the rotating screw. As the material approaches the die, there usually is an increase in pressure and temperature. At the entrance to the die plate, the temperatures and pressures are sufficient to force this extrudate through the die (Yacu, 1998).

Food extrusion usually involves cooking of the process materials. In the cooking process, sufficient energy, both thermal and mechanical, is imparted to gelatinize the starch and denature the protein. The majority of the food extrusion process includes extrudates that contain 8 to as much as 70% moisture, with most of the extrusion cooking processes occurring in the 10 to 30% moisture range.

The food extruder may be of single-screw or twin-screw design. The single-screw extruder is the most widely applied extrusion device in the food processing industry. The single-screw extruder also produces more tonnage of extruded products than any other extrusion processing method. The products produced by the single-screw extruder range from fully cooked, light density, corn snacks to dense, partially cooked and formed pasta.

The raw material characteristics, hardware selections, and effects of processing conditions will be examined in this chapter for various single-screw applications.

RAW MATERIAL CHARACTERISTICS AND SELECTION

Raw ingredient formulations, selection of processing equipment, and operating or processing conditions are independent regions of control that may be exercised in extrusion cooking. Although the control regions are independent, they are interrelated to the degree that discussions of one subject will usually involve the others (Rokey, 1994).

Ingredient selection has a tremendous impact on final product texture, uniformity, extrudability, nutritional quality, economic viability, and ability to accept coatings.

In general, extrusion converts cereal grain and protein blends into a dough. The starchy components gelatinize resulting in a substantial uptake of moisture and an increase in dough viscosity. Protein constituents may impact elasticity and gas-holding properties that are characteristic of hydrated and developed glutinous doughs. Other proteinaceous materials, those with low protein solubilities, may contribute less to the adhesive and stretchable functional properties.

Raw materials are selected primarily based on their nutritional contributions. Second, economics enters into the selection process. Third, the availability of the raw material becomes a factor.

When purchasing or selecting raw materials, a specification range based on desirable characteristics must be established. This should include the proximate analysis and other known qualities. However, sometimes desirable characteristics are only vaguely recognized and no test exists to monitor quality. Sometimes desirable characteristics are not even recognized.

One must recognize the existence of variabilities within a raw material due to influences such as the type of growing season of grains. Different types of grains and variation within animal species are reflected in the processability of raw materials.

Storage and processing of raw materials prior to extrusion greatly influence their reaction to heat, pressure, and shear. For example, cereal chemists recognize "after-ripening" factors which are biochemical changes that occur in grains during storage. In other words, grain that has been recently harvested extrudes much differently than grain that has been stored for six months. Whole grain is "alive" if it is sound and, therefore, changes with time.

To avoid surprises, develop historical databases that record raw material characteristics that correlate with good processing. Establishing a sample library of acceptable and unacceptable raw materials may be especially useful in maintaining a smooth running extruder.

Further considerations of recipe are discussed in the applications section of this chapter.

SELECTION OF HARDWARE COMPONENTS

Choosing the proper extruder configuration is critical to successful extrusion. The manufacturer of the extrusion equipment should be able to assist in tailoring configurations for processing a specific product.

There are many types of extruders, and each has a specific range of applications. An improper extruder selection for the specific application very rarely results in a smooth running process.

Once the proper extruder is selected, it must be assembled correctly and then adequately maintained. Training is important, and the supplier of extrusion equipment must be able and willing to provide this service.

Knowledge of what parts will wear and what their useful life will be will avoid costly and inconvenient shutdowns. Records are imperative in this endeavor and will greatly reduce the necessity of costly parts inventories.

The selection of processing equipment for a manufacturing plant is an important decision for the company that is considering food production. Equipment that will give the highest operating efficiency and most versatility at the lowest total cost should be chosen.

When sizing equipment for any plant, it is important to determine the rate or capacity at which the plant will be operating. However, the probability of expanding the plant in the future must also be considered. For example, it may cost very little extra to purchase a conveyor or storage bin that is capable of twice the presently needed capacity.

The extrusion system, whether a single-screw or corotating twin-screw configuration, must accomplish a number of phenomenon in a very short time under controlled, continuous, steady state conditions. These phenomenon include tempering, feeding, mixing, cooking, cooling, and shaping. The pressure, temperature, moisture, and resulting viscosity of the extrudate are affected by the system configuration and processing conditions.

Selection of the proper system configuration includes choosing from the following hardware components (see Figure 1):

a. Feed delivery system
b. Tempering or preconditioning phase
c. Extruder barrel components
d. Die and knife configurations

BIN

Ingredients, pre-ground and mixed to the appropriate recipe, are delivered to the holding bin above the extruder. This holding bin is of adequate volume to support the extruder operation for a minimum of five minutes. This minimum material requirement provides a buffer time for the operator and automatic control network to shut the extruder down in an orderly manner should there be a loss of raw material feed to the extruder. This holding bin is typically operated between high and low

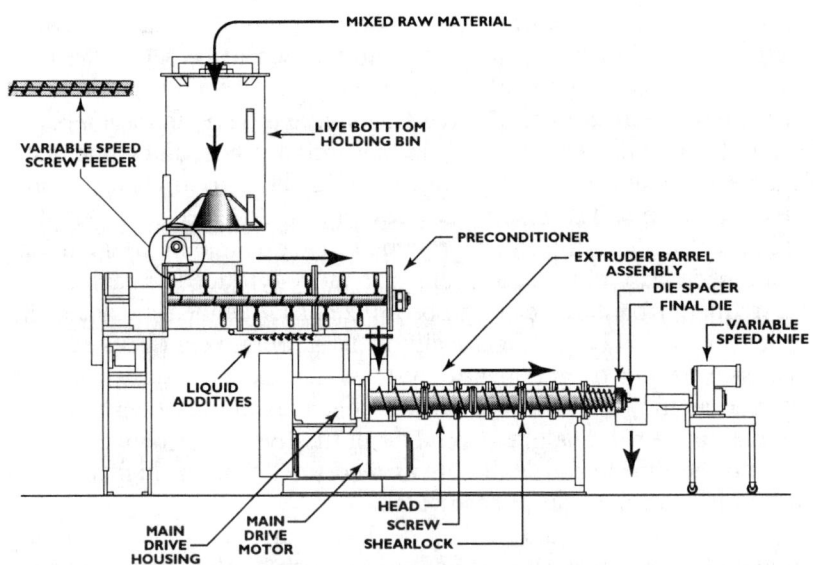

Figure 1 Cutaway view of single-screw extrusion cooker.

level limits. This bin must also insure that the extruder receives a continuous supply of material in a uniform manner, therefore, the bin is typically equipped with a device to prevent bridging of the dry mix.

FEEDING DEVICES

Feed delivery systems are an essential part of an extrusion operation. Consistent and uniform feeding of ingredients is necessary for the consistent and uniform operation of an extruder.

Hoppers or bins are an integral part of a feeding device and are used to hold the dry ingredients above the feeders. The feed delivery system must be able to uniformly feed a dry and/or liquid ingredient or blend of ingredients. Generally, when the added fat content of a raw formulation exceeds 12%, that portion of the fat above the 12% level should be introduced into the extrusion system in a separate ingredient stream. The dry feed portion is delivered to the extrusion system through a specialized metering device capable of providing uniform flow at any desired extrusion rate.

The extrusion rate of food extruders is typically controlled by a feeder screw or other metering device. The output of the single-screw cooking extruder is, therefore, independent of screw speed. The feeder screw, which delivers dry mix from the extruder bin, must be equipped with a variable speed drive to allow the screw speed to be set to the rotational speed necessary to deliver dry feed at the desired extrusion rate.

Dry ingredients are usually free-flowing, and there are a number of capable feeding devices that vary in their relative cost and complexity. Variable speed augers or screw conveyors can be used to volumetrically meter ingredients. These same devices can be designed and manufactured to act as loss-in-weight (gravimetric) feed systems by mounting the bin/feeder assembly on load cells. Vibratory-type feeders having variable frequency or stroke can also be utilized as gravimetric devices. Ingredients can also be gravimetrically metered with weigh belts.

Loss-in-weight or gravimetric systems are more expensive but are preferred as they are not influenced by changes in bulk density of the raw materials or the height of product in the live bin. Volumetric systems, on the other hand, will deliver feed rates that are higher with a full bin due to the material headweight.

Slurry tanks and liquid feeding devices (pumps) are utilized to accomplish uniform metering of liquid ingredients. The slurry tanks are often jacketed for heating or cooling and are equipped with agitators as required. Positive displacement metering pumps deliver metered liquids at constant rates by varying length of stroke or speed of rotation. Slurries or liquids can be premixed with dry ingredients but are preferably injected into preconditioning devices or the extruder barrel.

Preconditioning devices, the subject of a separate chapter in this book, will not be covered in this section.

EXTRUDER DRIVE

The extruder drive usually consists of an electric motor, a speed reduction system, a torque transfer system, and a bearing support mechanism. The electric motor size is determined by the product application. Low-speed forming extruders may require as little as ten kilowatt hours per ton of throughput, while high-shear stress applications (light density, expanded snacks) may require as much as 120 kWh per ton of throughput.

Speed reduction and torque transfer are accomplished either through the use of V-belts or gear reduction. V-belt drives have the advantages of less expense in the initial equipment installation, quieter operation, and a provision for torque overload protection. Gear drives are noisier than belts and require torque limiting devices. Some high-speed extrusion systems have the main drive motor directly coupled to the extruder shaft with some type of torque limiting device to protect the extruder against high loads.

The bearing systems on most single-screw extruder shafts are relatively simple. Most single-screw extruders in the food industry with installed motor power above 75 kW utilize a three-bearing arrangement. In this arrangement, the two outboard bearings support the radial load of the extruder shaft. A third bearing is added to absorb the thrust load from the extruder screw.

EXTRUDER BARREL

The extruder barrel is that portion that most often comes to mind when discussing the hardware components of an extruder. The extruder barrel assembly consists of a rotating extruder shaft and elements (maybe segmented screws and shearlocks), a stationary barrel housing (maybe segment sections), and a die and knife assembly. The length-to-diameter ratio of the extruder barrel can be varied as can the actual geometrical design of the individual components (screws, shearlocks, head sections, etc.).

The available screw elements vary depending upon the manufacturer and the application. In addition to simply transporting the material from the inlet to the die, screw geometry can influence mixing, kneading, and heat and pressure development. Common single-screw configurations are shown in Figure 2.

The design that includes variable pitch, constant depth, increasing root diameter, increasing number of flights, shearlocks, and decreasing diameter is most frequently employed in the food industry.

Single-screw extruders can categorize the movement and transformation of material within the extruder into three sections: feeding, kneading or transition, and final cooking zones (Figure 3) (Mercier et al., 1989).

The extruder rotating screw and shearlock elements sequentially convey and heat the material through mechanical energy dissipation.

The preconditioned or dry material enters the extruder barrel inlet with a typical density of 450–650 g/L. The inlet screw flighting is usually very deep and has a long pitch with flighting profiles that are nearly vertical to maximize transport (Figure 4).

As the material is conveyed into the kneading zone, its density may increase because of added water and steam. The screw pitch decreases and the flight angle may also decrease to accomplish more mixing in this area. The reduced slip at the barrel wall prevents the food material from turning with the screw, and this is referred to as "drag flow" (Figure 5) (Miller, 1990). A continuous screw channel serves as a path for "pressure-induced flow" as the pressure behind the die is usually much higher than that at the inlet. "Leakage flow" also occurs in the

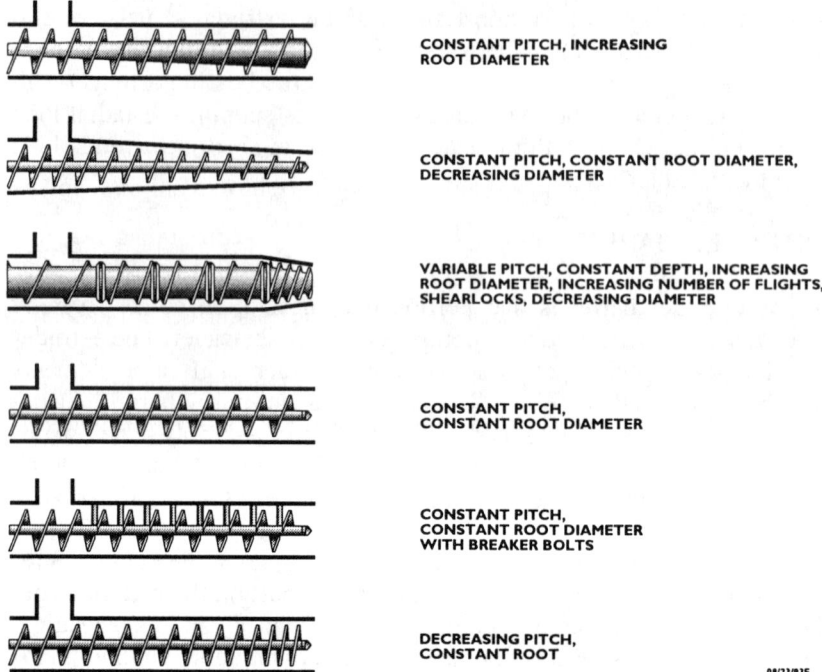

Figure 2 Common screw configurations for single-screw extruders.

clearance between the screw tip and the barrel wall. The flight of the screw may be interrupted in this area to further increase mixing via leakage flow. Energy inputs and moisture levels at this stage are usually present in sufficient amounts so that the material approaches its melt transition temperature. The material will exhibit a rubbery texture similar to a very warm dough.

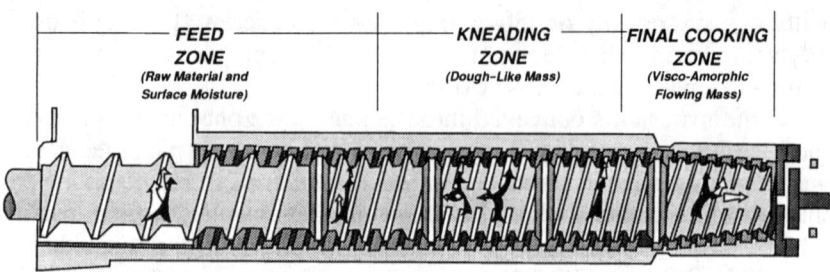

Figure 3 Extrudate transformation for medium- and high-shear single-screw cooking extruder.

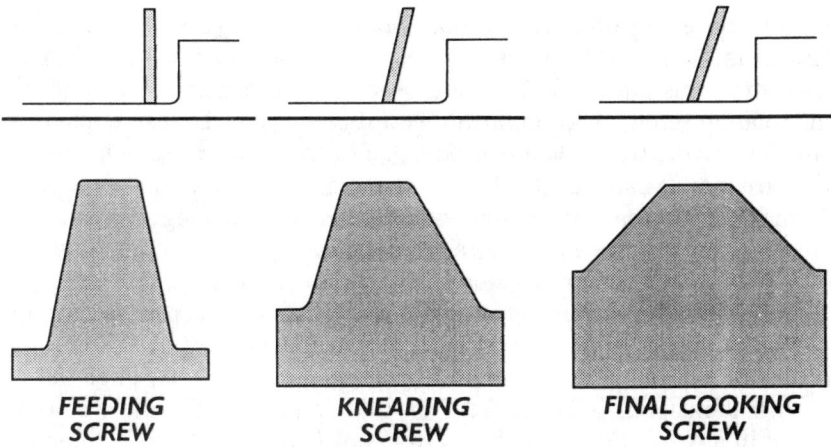

Figure 4 Screw flight profiles.

In the final cooking zone, density is further increased as the combination of thermal and mechanical energy inputs plasticizes the material above its melt transition temperature. The screw flight in the final cooking zone (Figure 4) is typically shallow, has a short pitch, and has flight profiles that are relatively flat to increase leakage flow and decrease conveying capacity. The final screw element is often conical-shaped (see Figure 3) which reduces volumetric displacement.

The extruder barrel wall may be designed to enhance certain cook-

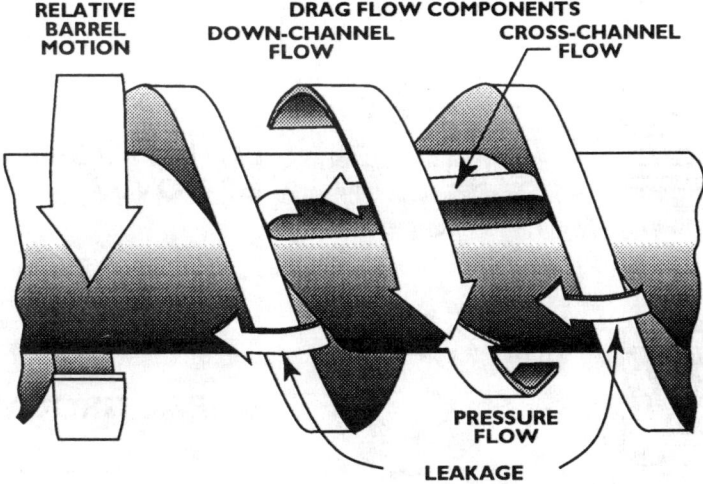

Figure 5 Flow components in the single-screw extruder.

ing or conveying processes. Smooth bore extruder barrels are not desirable in food extrusion applications due to the sticky nature of most food materials during the cooking process. The material must slip on the rotating screw to accomplish forward conveying. To insure that this slip occurs, the barrel wall is grooved. The axial groove design existed in extruders as early as the 1950s. These grooves have vital functions in many extrusion applications of avoiding or reducing slip and increasing throughput. Typical groove designs will either be longitudinal (straight-ribbed) or helical (spiral-ribbed) (see Figure 6).

Longitudinal grooves permit greater leakage flow across the top of the screw flight. Spiral-ribbed barrels also increase leakage flow, but act as an extension of the screw flight because their helix direction is such that material is conveyed forward. Spiral-ribbed barrels can be found in any location along the barrel configuration as their net contribution is increased conveyance. On the other hand, straight-ribbed barrel sections can increase cook on the extrudate especially if used where the barrel is filled with material.

The extruder barrel is usually segmented, and these individual seg-

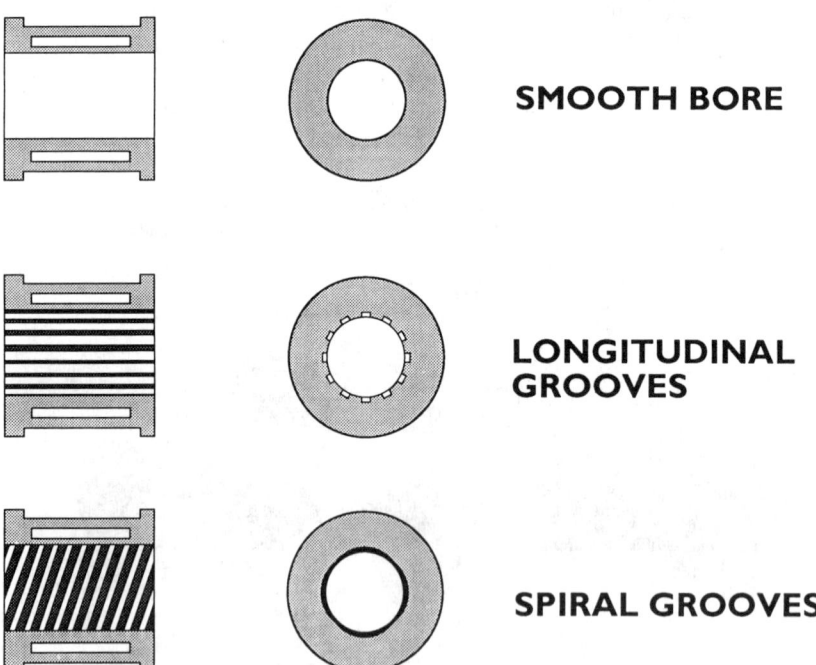

Figure 6 Typical single-screw extruder barrel wall configurations.

ments are jacketed to allow for individual zones of temperature control. A cooling medium such as water is usually circulated through the jackets in the feed zone to enhance feeding characteristics and to reduce material moisture from vaporizing. The kneading and final cooking zones are usually operated at higher temperatures. Steam at pressures of 4–6 bar is the most common heating medium, although hot water, thermal oils, and electric heaters are employed. Heat transfer into or out of the product through this means is relatively small compared to the overall energy supplied to the process. Short residence time in the barrel, incomplete barrel fill, low surface area to volume rations with larger diameter extruders, and low heat transfer coefficients of the barrel wall are responsible for this insignificance.

Die and knife technology for singe-screw extruders is similar to that for twin-screw extruders. The die plate serves as a restriction device at the end of the barrel which can control barrel fill, pressure, and temperature to some extent. The actual die openings determine the size and shape of the extrudate based on die design, extrusion configuration, processing conditions, and recipe.

PROCESSING CONDITIONS

Familiarity with the extrusion properties of ingredients and the interaction of such equipment parameters as screw speed and hardware design can allow a general classification of all extruders into three categories. These three categories rank extruders by relative shear stress and product categories (Table 1).

Low-shear stress extruders, often referred to as forming extruders, are used to densify material that is generally high in moisture. An example of products produced by food extruders in this category are pastas. Slow-speed extruders that have a long length-to-diameter ratio impart low levels of mechanical energy per unit of throughput.

Most pet food and aquatic feeds, along with textured vegetable proteins and breadings, fall into the medium-shear stress category. Note that moisture levels of the product are lower than the low-shear stress products, and that mechanical energy inputs are higher.

Highly expanded products comprise the high-shear stress category. Extrusion speeds and mechanical energy inputs are high, while product moisture levels and bulk density are low. Extruders of this type will have the shortest length-to-diameter ratio of the three classifications.

Once the hardware is selected based on the above discussion, the extrusion system can be operated under various processing conditions to achieve a multitude of final product qualities. Following is a list of independent variables that an extruder operator can directly manipulate:

TABLE 1. Single-Screw Extruder Classification.

	Lower Shear	Medium Shear	High Shear
Product Moisture %	25–75	15–30	5–8
Product Density (grams/L)	320–800	160–510	32–200
Maximum Barrel Temperature (°C)	20–65	55–145	110–180
Maximum Barrel Pressure (kg/cm2) (kPa)	6–63 588–6,178	21–42 2,059–4,119	42–84 4,119–8,238
D_s Screw Diameter h Channel Depth	3–5.3	5.0–8.5	8.0–18.0
Parallel Flow Channels (n)	1	2	2 or 3
Screw Speed (rpm)	less than 100	greater than 100	greater than 100
Energy Conversion (kw/kg)	0.01–0.04	0.02–0.08	0.10–0.16
Typical Products	Pasta Products Third-Generation Snacks Meat Products Gums	Textured Soy Breadings Expanded Pet Food Semi-moist Pet Food	Snacks Breakfast Cereals Breading (Croutons) Thin Boiling Starches

(1) Incoming recipe (the actual recipe, particle size, and moisture and temperature resulting from preconditioning)
(2) Rate (the rate at which the recipe is introduced into the extruder)
(3) Percent steam addition (steam at 6–9 bar can be injected directly into the material in the extruder barrel)
(4) Percent water addition (water at various temperatures and 3 bar can be injected directly into the material in the extruder barrel)
(5) Percent liquid addition (other liquids and/or gasses can also be introduced into the extruder barrel)
(6) Extruder and die configuration
(7) Temperature and flow rate of thermal fluid to barrel jackets
(8) Extruder speed (requires a variable speed drive)

When changes from the above list are made, they will in turn affect other operating variables (referred to as dependent variables):

(1) Material retention time in the extruder barrel
(2) Product temperature in the extruder barrel
(3) Product moisture in the extruder barrel
(4) Pressure in the extruder barrel

To reduce retention time in the barrel of the extruder, the feed rate of the incoming recipe should be increased. Turning the extruder at higher speeds and reducing the length-to-diameter ratio of the extruder barrel will also decrease product retention times.

Product temperature can be increased by increasing steam addition levels, using aggressive extruder configurations that maximize mechanical energy inputs, increasing the temperature of the thermal fluid circulating in the barrel jackets, and increasing screw speeds.

Every recipe has its own characteristics that also influence how a given extruder must be operated to attain desired product attributes. Strahm (1998) suggested that the principles of polymer science can be applied to recipe biopolymers such as protein and starch. This work indicated that total energy input (yielding a given product temperature in the extruder barrel) plotted against product moisture can yield "maps" to describe typical extrusion processes (Figure 7).

Note that the raw recipe starts at a point below the glass transition curve and that the plasticizing effect of preconditioning yields a soft, moistened material as it enters the extruder barrel. Further heating and moistening by thermal and mechanical energy input follows preconditioning. Moisture loss at the die cools and dries the product leaving a flexible, moist product that is then dried and cooled to produce a crunchy (glassy) final product.

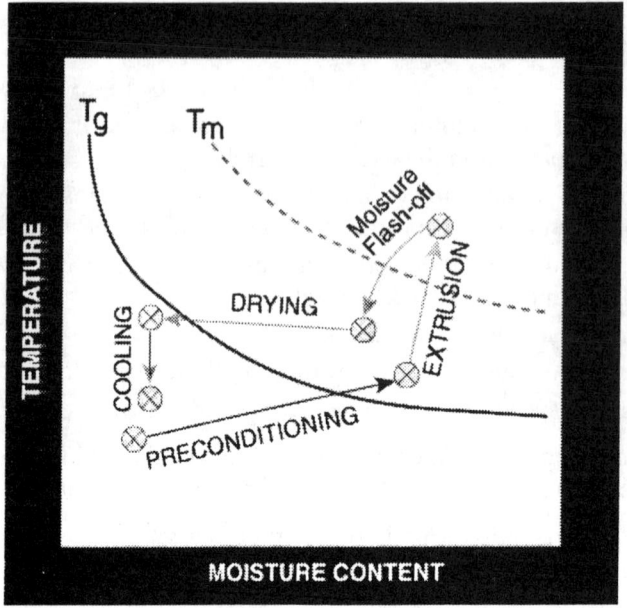

Figure 7 Glass and melt transition curves for a "typical" extrusion process.

A summary of typical parameters employed in various cooking processes appears in Table 2 (Rokey, 1995).

Note the broad range of operating parameters available in single-screw extruder systems. Typical production capacities and costs are shown in Table 3 for product types produced from the medium-shear stress category.

TABLE 2. **Typical Process Parameters.**

Process	Temp. °C	Max. BBL Pressure	Moisture %	Max. Fat%	Cook* %
Pellet Press	60–100		12–18	12	15–30
Expander/Pellet Press	90–130	35–40	12–18	12	20–55
Dry Extrusion	110–140	40–65	12–18	12**	60–90
Wet Extrusion					
Single-Screw	80–140	15–30	15–35	22	80–100
Twin-Screw	60–160	15–40	10–45	27	80–100

*Cook is starch gelatinization measured by enzyme susceptibility
**Dry extrusion successfully processes full-fat soy (18–20% fat) and other ingredients where final product durability is not a concern

TABLE 3. Typical Production Capacities and Costs.

Process	Capacity Range (ton/hour)	Ave. Life/ Major Wear Components (hours)	Ave. Wear Cost ($/ton) Full-Fat Soy	Ave. Wear Cost ($/ton) Complete Diet
Pellet Press	2–60	2,300	N/A	1.23
Expander/Pellet Press	2–40	700*	0.6**	1.14
Dry Extrusion	0.5–2	1,000	1.0	1.78
Wet Extrusion				
Single-Screw	1–22	5,000	0.50	0.89
Twin-Screw	1–14	5,000	1.85	2.01

*Reported 9,800 tons before replacement at a rate of 14 tons/hour
**Expander only—pellet press not used
N/A = Not Applicable

APPLICATIONS

HIGH-SHEAR STRESS PRODUCTS

Extrusion of products from high-shear stress processes will be examined first. Direct-expanded, light density snacks, starches, breadings, and breakfast cereals are examples of these types of products.

The most popular and successful extruders in the snack food production plant are single-screw extruders.

Snack extruders employ a number of phenomena in a short time under controlled, continuous, and steady state operating conditions. These phenomena may include heating, cooling, conveying, feeding, compressing, reacting, mixing, melting, cooking, texturing, and shaping. Energy sources to provide heat for cooking, texturing, or melting may include the following:

- *conversion:* Mechanical energy is dissipated into the extruder in the form of heat caused by shear or friction generated by pumping or con-

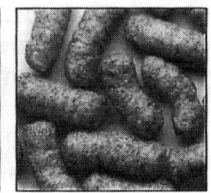

Photo 1 Direct expanded snacks.

veying inefficiencies. For high-shear extruders, it is possible that 100% of the heat may be generated through shear. This is referred to as an adiabatic extruder.
- *conduction:* Heat is transferred through the barrel of the extruder from a thermal fluid or other external source.
- *convection:* Heat is transferred to the extrudate by direct contact such as steam injection.

Each energy source has advantages and disadvantages. In most cases, a combination of the three is most practical.

The feeding zone of the extruder is the area where the low density discrete particles of raw material are introduced into the extruder barrel inlet. This low density, often preconditioned material is then transported into the interior of the extruder barrel. The flow channel of the screw is typically filled with low density material in this zone (Figure 3).

Density is low because of air trapped in the incoming granular raw material. The incoming material is compressed slightly in this zone and the air is expelled. Water is typically injected into the extruder barrel in the feeding zone to facilitate textural development and viscosity development and to enhance conductive heat transfer.

The kneading zone of the cooking extruder continues the compression started in the feeding zone, and the flow channels of the extruder achieve a higher degree of fill as a result of the compression of the extrudate. Within this zone of the extruder barrel, the extrudate begins to lose its granular definition, extrudate density begins to increase and pressure begins to develop in the extruder barrel. Shear begins to play a dominant role in the kneading zone because of the barrel fill condition that exists here. Barrel pressure is modest in the early part of the kneading zone. This modest barrel pressure permits, when desired, the injection of steam at a pressure between 5 and 10 atmosphere. When used, steam infuses thermal energy and moisture into the extrudate.

Within the kneading zone of the extruder barrel, the discrete particles of material begin to agglomerate due to the increasing temperature resulting from conduction, direct steam injection, and viscous energy dissipation.

The extrudate begins to form a more integral flowing dough mass as it moves through the kneading zone and typically reaches its maximum compaction. Shear in this area of the extruder barrel is moderate, and the extrudate temperature continues to increase.

The final cooking zone of the extruder barrel is where the mass becomes amorphous and texturized. Temperature and pressure typically increase most rapidly in this region as shear rates are highest because of the extruder screw configuration and maximum compression of the extrudate. Pressure, temperature, and resulting fluid viscosity are such that the extrudate is forced from the extruder creating the desired final product texture, density, color, and functional properties.

Single-screw extruders have some process limitations in the food processing industry. The most obvious limitation centers around the ability of single-screw extruders to transport sticky and/or gummy raw materials. Single-screw cooking extruders are further limited in their ability to process materials which become extremely sticky and/or gummy during heating and compression or which react adversely to the shear environment of the cooking extruder. From a process application and quality control point of view, the single-screw extruder may be limited in its ability to maintain process stability.

Process instability in single-screw extruders generally manifests itself in product surging, i.e., non-steady-state flow of the extruder die.

The majority of extruded snacks on the market fall into this category which may be classified as fried snacks or baked snacks.

Properly selected cornmeal is fed into an extruder with a feeding device that delivers the meal at a constant rate. The meal is exposed to moisture, heat, and pressure as it is transported through the extruder into the extruder die. As the material exits the extruder die, it expands due to pressure release and is sized to the proper length with a rotating knife to produce a collet or ball shape. These collets are then dried and coated with a seasoning (Figure 8).

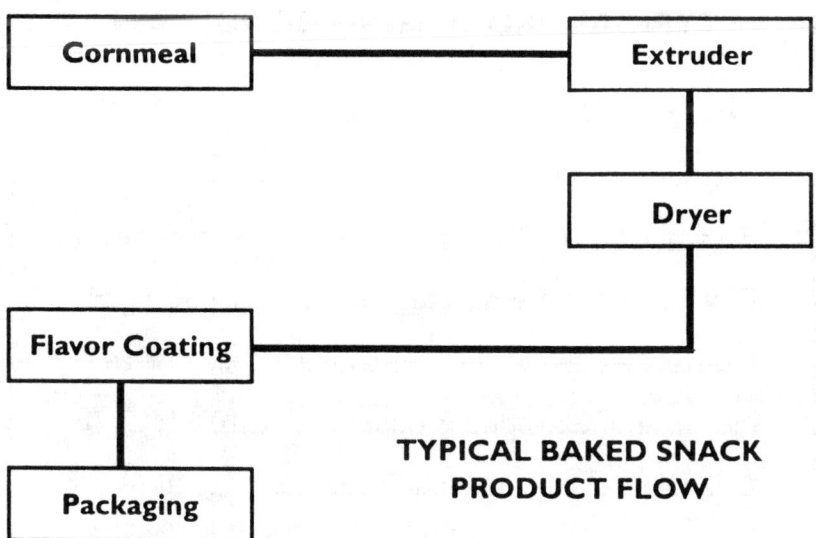

Figure 8 Typical direct-expanded product flow.

Extruder barrel lengths are usually less than 10:1 length:diameter ratio. Typical processing conditions are found in Figure 9.

The specifications or grade of cornmeal used tend to influence the texture and mouthfeel of the final snack product. Many grades or granulation of degermed cornmeal are available. For example, if a finer texture, cell structure, or softer bite for a snack is desired, a snack producer may wish to use raw materials having a typical analysis as indicated in Table 4. Conversely, some consumers prefer extruded snacks that exhibit a crunchier texture with a slightly larger cell structure. For this type of texture, a coarser granulation of degermed cornmeal is desired as indicated in Table 5.

Premoistening of raw cornmeal when using a course granulation may be required to eliminate grittiness in the final product. A large amount of flour in the degermed cornmeal is undesirable, especially when using a single-screw extruder. Flour tends to segregate in the feeding system and in the inlet portion of the extruder barrel.

When water is added in the extruder barrel, flour absorbs the water quickly, leaving less moisture available for the coarse particles. This phenomenon may disrupt the product flow and cooking of the extrudate which causes fluctuations in final product quality. Table 6 shows a troubleshooting guide for expanded snacks.

Moisture levels of the extruded snacks are normally between 8 and

TYPICAL PROCESSING CONDITIONS

Extruder RPM 300-400

Extruder Barrel Temperature .... 120-160°

Extruder Barrel Pressure .... 70-150 Atmospheres

Dry Cornmeal Feed Rate .... 450 Kg/Hr

Moisture Added into the Extruder .... 13 Kg/Hr

Moisture of Extruded Product .... 8-10% MCWB

Bulk Density of Extruded Product .... 48-64 g/l

Figure 9 Typical processing conditions.

TABLE 4. **Specifications for Softer Textured Degerminated Cornmeal.**

Analysis Performed	Typical Analysis	Analysis Range	
		Minimum	Maximum
% Moisture	11.0	10.0	12.0
% Protein	8.0	7.5	9.0
% Oil	0.8	0.75	0.9
% Fiber	0.4	0.3	0.5
% Ash	0.3	0.1	0.5
% Protein	8.0	7.5	9.0
% Nitrogen Free Extract	79.7		
Bulk Density lb/ft^3 Loose	44.0		
% Granulation on U.S. Sieve			
On 30	0.1	0.0	0.5
On 40	0.4	0.0	1.0
On 60	80.0	70.0	90.0
On 80	15.0	10.0	20.0
Through 80	4.5	1.0	7.0

TABLE 5. **Specifications for Crunchier Textured Degerminated Cornmeal.**

Analysis Performed	Typical Analysis	Analysis Range	
		Minimum	Maximum
% Moisture	11.0	11.0	12.0
% Protein	7.5	6.5	8.5
% Oil	0.7	0.4	1.0
% Fiber	0.5	0.3	0.6
% Ash	0.4	0.3	0.6
% Protein	8.0	7.5	9.0
Bulk Density lb/ft^3 Loose	44.0		
% Granulation on U.S. Sieve			
On 10	0.0	0.0	0.1
On 14	2.5	0.0	5.0
On 20	65.0	60.0	75.0
On 40	31.0	25.0	40.0
Through 40	0.5	0.0	2.0

TABLE 6. Troubleshooting Guide—Expanded Snacks

Problem	Extruder Solution	Mechanical Solution
Density Too Light	(1) Add water to the extruder barrel (2) Cool down extruder heads (3) Reduce dry feed rate based on open area in the die	(1) Barrel temperature must be constant (2) Change water injection location (3) Increase open area in the backup die or the final die (4) Increase the clearance between the final die and the backup die
Density Too Heavy	(1) Reduce water to the extruder barrel (2) Heat the extruder heads—normal range 120–165°C (3) Increase rate	(1) Decrease open area in the backup die or the final die (2) Decrease clearance between final die and backup die (3) Add 0.25–0.50% baking soda
Rough Surface of Product	(1) Extruder heads may be too hot, try cooling the heads (2) Extruder water injection too low	(1) Increase open area in final die (2) Check wear of die screw and sleeves
Texture Too Hard, Large Cell Structure, Crunchy Texture	(1) Reduce water to the extruder barrel (2) Heat extruder heads	(1) Too much open area in final die (2) Reduce size in backup die or clearance between final die and backup die
Hard Spots	(1) Heat extruder heads (2) Premoisten meal (3) Reduce size of raw material particle size	(1) Check extruder for proper configuration (2) Too much clearance in die (3) Measure extruder parts for wear

TABLE 6. (*continued*)

Problem	Extruder Solution	Mechanical Solution
Surging or Extruder Instability.	(1) Check for buildup of raw material in the inlet head (2) Temperature too high in extruder—cool heads	(1) Measure extruder barrel for wear (2) Product buildup in die from improper start-up procedure (3) Product buildup on cone screw (4) Barrel temperature must be constant (5) Change water injection location
Knife Blades, Tails, Uneven Shapes		(1) Make sure knives are sharp and set tight (2) Make sure the proper amount of knives is being used for rate and shape
Extruder Start-up	(1) Heat extruder cone heads (2) Start extruder water injection 40% higher than running condition (3) Start raw materials lower than full rate (4) Increase feed rate after extrudate is flowing from extruder die (5) Adjust head temperature to cool if necessary (6) Reduce extruder water to get desired product	
Shutdown Procedure	(1) Cool down extruder barrel (2) Add extruder water (3) Decrease dry feed rate (4) Product becomes wet— stop feed (5) Stop water (6) Remove die (7) Flush extruder	

12% and require additional drying to impart the desired final product texture and mouthfeel. Temperatures of 150°C and retention times of 4–6 minutes during drying are required to achieve the desired low moistures of 1–2%. Before coating, the extruded products with seasonings and the fines that are generated during conveying and drying are separated. Fines tend to absorb the flavor and oil coatings causing agglomerated balls. Typical seasoning recipes for coating direct-expanded corn snacks are similar to the following:

INGREDIENT	PERCENT
Extruded Corn Snack Base	60.8
Coconut Oil	28.0
Cheddar Cheese Powder	7.0
Acid Whey Powder	3.5
Cheese Flavor	0.5
Salt	0.2

MEDIUM-SHEAR STRESS PRODUCTS

Examples of products in this category include textured vegetable proteins and pet or aquatic feeds.

The single-screw cooking extruder has been the "heartbeat" of the dry expanded pet food and aquatic food industries in the U.S. for over thirty-five years. They represent the world's largest volume product produced by extrusion cooking.

Because pet food holds the position as the largest volume of extruded goods, futuristic equipment manufacturers are investing heavily in research to better their equipment and services.

Single-screw cooking extruders continue to produce a major portion, perhaps as much as 90%, of dry expanded pet food. New screw and barrel geometries have been developed and are being installed in new plants and in plant expansions. State-of-the-art hardware is also being retrofitted, where possible and practical, to existing production equipment.

Photo 2 Pet and aquatic foods are examples of medium-shear stress products.

The preset screw and barrel configurations represent many years of analytical design, research and comprehensive testing. A better understanding of the machine/materials interaction occurring inside the extruder barrel has led to the development of screw and barrel geometries for single-screw extruders that are more efficient in converting mechanical energy to heat through friction. These screws have increased volumetric capacity permitting higher levels of steam injection into the extruder barrel, further increasing throughput and energy efficiency. Medium-shear stress products require length:diameter ratios of 6:1 to 18:1. Moistures in the extruder are 16 to 30% and extruder screw speeds range from 400 to 1,000 RPM, depending on the extruder diameter. Most importantly for process and final product quality control, the pumping and conveying actions of present-day single-screw extruders have been vastly improved by redesigning the screw flight profile and groove geometry in the barrel wall. For many product applications, single-screw extruders exhibit die pressure and flow stabilities matching those of twin-screw cooking extruders. Uniform flow within single-screw extruder barrels yields die pressure stability and constant uniform extrudate delivery to the die. This results in minimal variations in the degree of cook in the product, while virtually no variations are seen in product shape and size.

As new technology has unfolded in structural design and metallurgy, single-screw systems have screws and barrel sleeves which have historically demonstrated usable lines of 6,000 to 7,000 hours of operation. Single-screw throughput capacities have exceeded 12 to 22 ton per hour capacity on the large extruder systems.

It may be desirable in single-screw systems to produce a dense pellet that is fully cooked. Most of the sinking aquatic feeds fall into this category. A nonexpanded, fully cooked product is particularly difficult with formulations that are high in starch content and contain less than 12% fat. To reduce product temperatures, moisture, and expansion, the extrudate can be subjected to a decompression step prior to extrusion through the final die. Extrusion systems with and without vented head configurations are found in Figures 10 and 11.

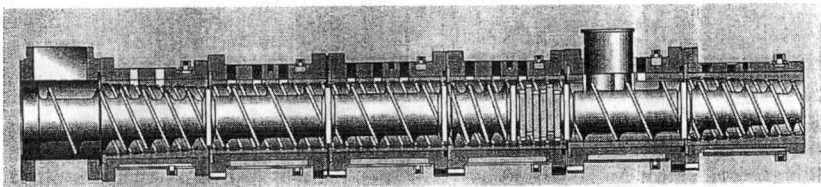

Figure 10 Extruder barrel assembly for the production of sinking aquatic feeds.

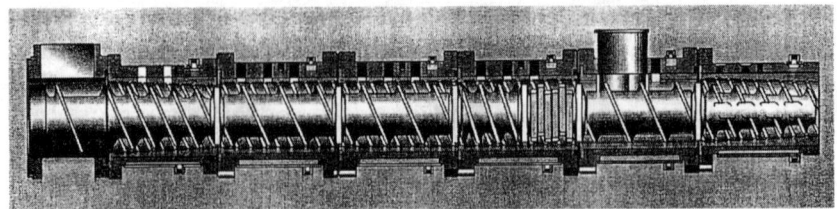

Figure 11 Extruder barrel assembly for the production of floating aquatic feeds.

The extruder configurations for floating and sinking aquatic feeds appear quite similar. This is by design to minimize the necessary component changes when switching from one product to the other. However, the configuration for sinking aquatic feeds has an open vent that effectively reduces expansion of the extrudate. The fine cone screw segment in the configuration for floating aquatic feed is an interrupted-flight design for increasing leakage flow and mechanical energy input to expand the product as it emerges into atmospheric conditions after the die. The vented head apparatus is also closed during the processing of floating diets to maintain pressure in the extruder barrel.

The vented head also provides an opportunity to introduce dry or semidry ingredients into the latter portion of the extruder barrel after a significant amount of cooking has occurred.

LOW-SHEAR STRESS PRODUCTS

Pasta products typify this category of products.

This family of products is to be distinguished from the family of quick-cooking or instant pastas that often employ twin-screw extrusion systems and portions of a process that could be categorized as medium shear stress conditions.

Semolina, the purified middlings of durum wheat, is the preferred raw material for the production of pastas or macaroni (Harper, 1981). The uniform size of semolina makes it easier to mix with water, and

Photo 3 Pasta products.

the amber color is a desirable quality. If the semolina contains large amounts of flour, the water will be preferentially absorbed by the flour, and the coarse particles may not be hydrated, resulting in white specks in the dried product. To manufacture pastas, water is blended with durum semolina or flour to create a resulting dough of 31% moisture.

The doughs are blended in continuous mixers where dry ingredient feeders and metering pumps deliver the correct ratio of the recipe. The dough in pasta extruders is subjected to a vacuum of -63 kPa to -80 kPa which removes entrapped air before extrusion. Pasta that has been extruded after vacuum treatment has a smooth surface, a bright and clear color, and a better texture. The entrapped air appears as bubbles in the pasta which reduces transparency and detracts from the appearance and color of the final product. Vacuum treatment usually involves drawing a vacuum on the dough before it enters the screw. Often, the entire mixing chamber is sealed at the inlet and discharge and the dough moves continuously through rotary valves to maintain the seal.

The purpose of the actual extruder screw in the pasta extruder is to knead the semolina/water dough and deliver it uniformly to the die. Heating of the dough is undesirable because these pasta products are uncooked until they are placed in boiling water during their final preparation. To meet the special requirements of this low shear stress application, screw and barrel designs have unique features.

Pasta screws are usually deep-flighted with constant flight height and uniform pitch the entire length. Mixing within the flights is minimal, and it is often necessary to interrupt the flighting near the discharge end of the screw to assure that the dough is uniformly mixed as it nears the die. A breaker plate is another technique that might be employed to ensure uniformity of the final product. The breaker plate is usually part of the die assembly and consists of a heavy, perforated metal plate. This plate can serve as a support for a fine screen upstream of the plate. This screen breaks the flow into very small streams that recombine after they flow through the plate to enhance mixing and to hold back small foreign pieces that could otherwise plug the die.

Low-temperature extrusion ($<50°C$) is desirable to produce superior quality in the final product. It is necessary to remove heat that is added during extrusion in the form of mechanical energy input. Cooling water is circulated through the jackets of the extruder barrel to remove this heat.

REFERENCES

Harper, J. M. 1981. "Macaroni extrusion." In: *Extrusion of Foods*, Vol. II, Boca Raton, FL: CRC Press, pp. 20–29.

Mercier, C., P. Linko, and J. M. Harper. 1989. *Extrusion Cooking*. American Association of Cereal Chemists, St. Paul, MN, p. 3.

Miller, R. 1990. "Unit operations and equipment." In: *Breakfast Cereals and How They Are Made*, American Association of Cereal Chemists, St. Paul, MN, p. 145.

Rokey, G. 1994. "Better extrusion." Pet Food Industry. 36(4): 10.

Rokey, G. 1995. "Extrusion technology and nutritional implications." Presented at Expo Aviga, Barcelona, Spain.

Strahm, B. 1998. "Fundamentals of polymer science as an applied extrusion tool." Cereal Foods World. 43(8): 621–625.

Yacu, W. A. 1998. *An Overview of Single Screw Cooking Extruder, Food Extrusion Short Course*. The Center for Professional Advancement, New Brunswick, New Jersey.

Yacu, W. A. 1999. *Introduction and General Principles of Food Extrusion, Food Extrusion Short Course*. The Center for Professional Advancement, New Brunswick, New Jersey.

CHAPTER 3

Dry Extruders

NABIL W. SAID

THE objective of this chapter is to review the history of dry extrusion development, the principle of dry extrusion and its classification, and the different applications in the feed and food industry along with the latest technology of dry extruding-expelling oil from oilseeds.

The dry extrusion process was developed by Triple "F," Inc. in the mid-1960s to enable farmers and producers in the Midwestern U.S. to process whole soybeans that could be fed directly to livestock and poultry.

Some of the objectives of its development were the simplicity of the extruder operation, safety, the nutritional consequences of the process, the low capital investment and operational costs, and the versatility of the machine in processing other grains and complete feed.

In 1969, the first dry extruder was introduced, and the Insta-Pro division of Triple "F," Inc. was established. The extruder was given the name Insta-Pro, reflecting the instant protein processing that took place in a matter of seconds. The extruder and the process were patented thereafter.

THE PRINCIPLE OF DRY EXTRUSION

The term dry extrusion, as opposed to wet extrusion, refers to the fact that this type of extruder does not require an external source of heat or steam. All the cooking is accomplished by friction capitalizing on the inherent moisture and/or oil for providing lubrication. The dry extruder can process material bearing 8–22% moisture without the need for drying the extrudate. Within this range of moisture content, the extrudate will retain half of the original moisture of the starting material. The dry extruder can be configured to fully cook material containing up to 40% moisture, but only 10% point moisture will be lost, thus requiring drying of the extrudate from 30% to less than 12%.

The dry extruder is fitted with a water injection device that can be used in adding water directly into the barrel whenever it is needed. Recently, a preconditioner was provided as an option for adding a small amount of dry steam for applications that require a more uniform shape of formed products or for increasing the capacity and reducing the wear of the parts. Again, there is no need for drying since the moisture content of the extrudate does not exceed 12% in most circumstances.

CLASSIFICATION OF THE DRY EXTRUDER

The dry extruder is classified as a single-screw autogenous or adiabatic extruder that generates heat through friction. It is also considered a high-shear machine that can achieve high compression ratios, high temperatures, and various degrees of puffing.

It is also known as a high-temperature/short-time extruder (HT/ST), as temperatures up to 180°C can be achieved in less than 20 seconds, and as a low-cost extruder (LCE) due to its low capital cost.

COMPONENTS OF THE SINGLE-SCREW DRY EXTRUDER

The main part of the extruder, the barrel, is segmented into two to six chambers depending on the model and capacity of the extruder. The root of the screw in all models is constant in diameter, and compression results by changing the pitch of the worm flights. The degree of shear can be achieved by selecting the size of the steamlocks and the screw flight and adjusting the nose bullet and nose cone in the last chamber of the extruder barrel. The barrel wall and the steamlocks are grooved to allow more mixing and shearing.

The segmented screw worms and barrel sections offer considerable versatility in customizing screw barrel design and replacing worn parts.

The stock material can be fed to the extruder barrel through an over-the-top AC or DC controlled volumetric feeder with agitator for material that is uniform and free flowing. In the case of less uniform, hard-to-flow stock material, a screw side feeder is used to force the stock material directly into the inlet chamber of the barrel.

As the product enters the inlet chamber, it will be forced to move forward with the screw to the first steamlock. Because of the grooves in the wall and steamlocks, a gradual buildup in pressure is achieved to pass through the compression chambers. At the point the extrudate reaches the last chamber containing the nose bullet and the nose cone, the maximum pressure of about 40 atmosphere is achieved and the preset product temperature is fine-tuned by adjusting the nose bullet while the extruder is running with a constant shaft speed.

Products can be shaped by replacing the nose bullet with a flat bullet and attaching the needed auxiliary unit to the end of the barrel. An AC frequency drive unit, flaking head, cutter head, or a double die compression head and a wiper with variable speed can be mounted on any model extruder to perform the needed task in shaping or texturizing a particular product.

The smallest dry extruder available with a capacity of 600–800 pounds per hour requires a 50 horsepower electric motor. It can operate by power takeoff with a 100 HP tractor.

The larger models require 75 HP (1,300–2,000 lbs/hr), 125 HP (1,980–2,980 lbs/hr), and 350 HP (6,000–10,000 lbs/hr). Those capacities can be doubled when a steam preconditioner is fitted to the dry extruder (Figure 1).

The extruder drive is a single-ribbed, nonslip belt for all models except for the largest capacity model (350 HP) which is driven by an in-line gear box.

THE APPLICATION OF DRY EXTRUSION

Although the dry extruder was initially designed for the on-the-farm cooking of soybeans and cereal feed, today its utilization in numerous applications reflects its versatility and ability to obtain the needed objective in a cost-effective way. Below are some of the current applications of dry extruders:

a. Processing oilseed including soybeans, canola, rapeseed, cottonseed, sunflower seed, and peanuts
b. Producing complete feed such as pig starters, aquatic feed (floating or sinking fish feed), dog food, cat food, mink and fox feed, poultry feed, horse feed, and calf starter feed
c. Processing of high-moisture by-products of the animal, food, and marine industry into high-quality feed ingredients
d. Playing an integral part in the "Express" extruder/press trademark system of expelling oil from oilseed (Said, 1998) (The dry extruder's basic function is pretreatment of the seed prior to pressing the oil.

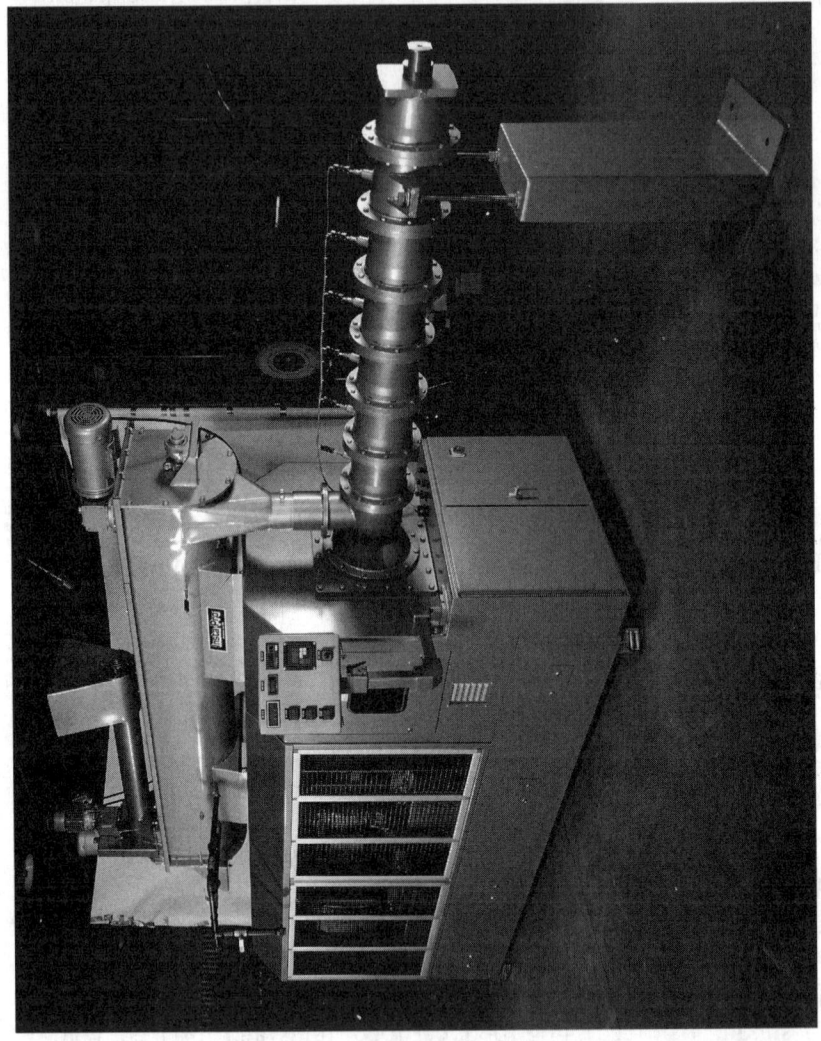

Figure 1 Insta-Pro Model 2500 extruder with the optional preconditioner for processing shaped products, complete feed, and whole soybeans.

The extruder cooks the oilseed, ruptures the cell, and partially dehydrates the extrudate prior to entering the press, thus increasing the rated capacity of the press and its efficiency.)

e. Stabilizing rice bran through the deactivation of the naturally occurring enzyme, lipase
f. Producing texturized plant protein and manufacturing cereal food
g. Acting as a reactor to detoxify peanuts or cottonseed through the reduction of aflatoxin from both seeds and the reduction of the free gossypol in cottonseed
h. Pasteurizing and/or sterilizing feed and food
i. Deactivating antinutritional factors present in legumes
j. Producing high nondegradable (bypass) protein for ruminant animals
k. Producing slow release ammonia products for the safe feeding of a nonprotein nitrogen to ruminant animals

NUTRITIONAL ADVANTAGES OF THE DRY EXTRUSION PROCESS

Since its introduction in the marketplace, the dry extruder has been studied and compared to other processing techniques relevant to the nutritional consequences of feeding the extruded product produced by this method to different specie. The nutritional advantages were reported in many scientific journals, at meetings, and in articles. Because of the extensive literature available on this subject, no attempt will be made here to fully report all the data, but the conclusions of the research will be highlighted.

EXTRUSION OF WHOLE SOYBEANS

Raw soybeans cannot be used effectively as an animal feed ingredient. There are many antinutritional factors in raw soybeans that can hinder the proper digestion of nutrients. Some of these antinutritional factors include the following:

a. *Trypsin and chymotrypsin inhibitors*: They interfere with the proper digestion of the protein. Kakade et al. (1973) estimated that nearly 40% of the reduction in growth performance of animals fed raw soybeans is due to the trypsin inhibitors. Clinically, hypertrophy of the pancreas occurs in these animals due to the effort of the pancreas to offset these factors by increasing the secretion of the enzymes trypsin and chymotrypsin (Mian and Garlich, 1985).

b. *Hemagglutinins (lectins)*: They interfere with the digestion of carbohydrates by interfering with the function of the enzyme amylase.
c. *Urease*: This enzyme is of importance in monogastric nutrition only as a guide for measuring the adequacy of processing. It is of some concern in diets for ruminants, as these diets quite often contain considerable amounts of urea. Urease can split urea into two molecules of ammonia which can be toxic to the animal.
d. *Allergenic factors*: Glycinin and beta conglycinin are the storage proteins in soybeans. They affect the integrity of the microvilli of the small intestine, thus resulting in malabsorption of nutrients and scouring in young animals.
e. *Lipase and lipoxygenases*: They result in peroxidation and beany flavor, respectively.
f. *Phytic acid*: This interferes with the proper absorption of minerals, particularly zinc.
g. *Isoflavones*: These possess estrogenic activity that may result in reproductive problems.

All but the last two antinutritional factors can be deactivated through proper heat processing such as occurs in extrusion. Because all of these antinutritional factors are proteins in nature, the type of heat process can also negatively affect the oilseed protein. Thus, caution must be exercised in order to produce a soy product that is high in digestible nutrients, especially energy and protein, while reducing the level and altering the structure of those antinutritional factors and rendering them inactive.

The high-temperature/short-time cooking provided by the dry extruder seems to accomplish this function better than any other process. The avoidance of further exposure to high temperatures through drying that may be necessary for the moist type of extrusion processing techniques could be the reason for maintaining the highest protein and amino acid digestibility.

Poultry Trials

In a broiler feeding trial, Cheva-Isakaru and Tangtaweeipat (1995) reported a superior destruction of the trypsin inhibitor factor by dry extrusion (an average of 11.2% and 74.5% higher than steaming and roasting, respectively). Broilers fed dry extruded whole soybeans had the best feed-to-gain ratio ($P < 0.05$) as compared to broilers fed soybean meal, steamed, and roasted full-fat soybeans.

Wiseman (1984) reported that both Apparent Metabolizable Energy (ME_A) and nitrogen retention as determined in two-and-a-half-week-old chicks, were the highest from soybeans extruded by dry extrusion (Insta-Pro extruder) as compared to soybeans processed by toasting, jet sploding, micronizing, wet extrusion, solvent extracted soybean meal plus added oil or raw soybeans.

Waldroup and Hazen (1988) fed soybean meal, roasted soybeans, dry extruded soybeans, and raw soybeans to laying hens in order to draw comparisons. Their results indicated that hens fed dry extruded soybeans achieved the highest egg production and the best feed conversion as compared to diets containing roasted soybeans, soybean meal, and raw soybean.

Stillborn et al. (1987) conducted experiments to compare the effects of three different processing techniques on broiler performance. They reported significantly higher body weight and better feed efficiency ($P < 0.05$) in broilers fed whole soybeans processed by dry extruder (Insta-Pro) compared with wet extruder or Gem Crop roaster.

Zhang et al. (1993) studied the effects of dry extrusion and expelling on the nutritional quality of conventional (CSB) and Kunitz trypsin inhibitor-free soybeans (KFSB) in chickens. Their results indicated that increasing the extrusion temperature of the CSB to 138°C and 154°C significantly increased the True Metabolizable Energy (TMEn) compared with CSB extruded at 104°C or 121°C (3,815, 3,936 vs 3,665, 3,678 kcal/kg DM, respectively). Extruding KFSB at 104°C, 121°C, and 138°C resulted in similar TMEn as that of CSB extruded at 138°C and 154°C. The extruded-expelled CSB meal had a TMEn value of 3265 kcal/kg DM. They reported that the amino acid digestibility of CSB did not differ from that of extruded KFSB. The expelling process had no significant effect on the amino acid digestibility values when compared to extrusion alone.

Swine Trials

Marty and Chavez (1993) reported significantly superior ($P < 0.05$) digestible energy and crude protein across all stages of pig growth when dry extruded whole soybeans were compared with micronized, jet sploded, roasted whole soybeans or soybean meal.

Kim et al. (1994a) conducted metabolism studies on nursery pigs to determine the effect of dry extrusion vs roasting on the nutritional value of two varieties of soybeans. They reported that nitrogen digestibility, biological value, percent nitrogen retention, gross energy digestibility, and metabolizable energy were greater for pigs fed extruded soybeans as compared to pigs fed roasted soybeans. They reported that the National Research Council (NRC) ME_A value of 1,664 kcal/lb for full-fat soybeans should be revised to indicate the type of processing used.

In another trial, Kim et al. (1994b) determined the ileal digestibility of nutrients in growing and finishing pigs from either dry extruded whole soybeans, roasted whole soybeans, or soybean meal. Their results indicated that digestibility of dry matter, gross energy, nitrogen, and various amino acids was superior for extruded soybeans as compared to roasted. Soybean meal nutrients digestibility was lower than that of dry extruded whole soybeans but was higher than that of roasted soybeans.

Hancock et al. (1991) reported that the improved feed efficiency for pigs fed dry extruded sorghum and soybeans indicates that the ME_A value for sorghum grain is dependent on the processing method and that the National Research Council (NRC) value for ME_A of heat-processed soybeans is probably too low, at least for dry extruded whole soybeans. Their data showed a 6% increase in dry matter digestibility and a 14% increase in nitrogen digestibility due to extrusion of soybeans and sorghum as compared with soybean meal-sorghum-soy oil non-extruded control diets.

Woodworth et al. (1998a) conducted an experiment to determine the apparent ileal digestibility of amino acids and digestible and metabolizable energy values of conventional soybean meal or dry extruded-expelled soybean meal for swine. Their results showed that dry extruded-expelled soybean meal with or without the hulls removed has a significantly higher ileal amino acid digestibility, crude protein digestibility, nutrient density, and metabolizable energy than solvent extracted soybean meal. In a second trial, Woodworth et al. (1998b) conducted a feeding trial utilizing the nutrient digestibility values obtained from the first trial. When formulating based on digestible nutrients, less extruded-expelled meal was required to match the digestible nutrient level in the complete diet as compared to solvent-extracted soybean meal and added soy oil or a commercial expelled meal (Soy Plus). Soy Plus fed pigs had significantly lower average daily gain and feed efficiency as compared to either extruded-expelled soybean meal or solvent-extracted soybean meal.

Ruminant Trials

Aldrich and Merchen (1995) studied the effects of heat treatment of whole soybean on protein digestion by ruminants. They reported that increasing the dry extruder temperature from 200°F to 320°F at 20°F increments resulted in a linear decrease of *in situ* degradation of soybean protein. As expected, raw soybean protein degraded very quickly. The rate of degradation for the raw, 200°F, 240°F, 260°F, 280°F, 300°F, and 320°F treatments were 84.1%, 45.7%, 40%, 40.9%, 48.6%, 36.7%, and 30.4%, respectively. In other words, extruded soybeans at 320°F had a bypass protein value of 69.6% compared to 15.9% for raw soybeans.

They determined the total amino acid digestibility before and after ruminal incubation utilizing the precision-fed cecectomized roosters assay. Their results showed that unincubated raw soybeans had a value of 68.5% compared to 87.7% for extruded soybeans at 320°F. Digestibility of residues of extruded soybeans averaged 90% at the different extrusion temperatures as compared to raw soybeans (82%).

Socha and Satter (1991) conducted a study to determine the production response of early lactating cows fed either solvent-extracted soybean meal, raw soybeans, dry extruded whole soybeans, or roasted soybeans with alfalfa silage as the sole forage source. They reported lower dry matter intake for cows fed raw or roasted soybean treatments. Cows fed dry extruded whole soybeans produced more milk, milk protein, and more 3.5% fat corrected milk than cows fed the other diets. Body weight changes and body condition scores did not differ among the treatments.

Smith et al. (1980) reported that early lactating cows fed extruded soybeans (7% dry matter basis) produced an average of 2.9 kg more milk per day than a control diet without extruded soybeans.

DRY EXTRUSION RECYCLING OF BY-PRODUCTS

Most by-products contain high moisture and cannot be effectively dried or dehydrated without adversely affecting their nutritional values. Processing such by-products through dry extrusion (Figure 2) was accomplished by means of blending those by-products with a local ingredient of choice to reduce the moisture content to a level suitable for dry extrusion. The blend is then fully cooked, sterilized, and partially dehydrated in a matter of seconds. The rupturing of cell walls that takes place frees the moisture from inside the cells, yielding an easily dehydrated final product. The moisture is then reduced by a thermal dryer to about 10% prior to cooling and storing the finished product.

This process has been tested by many universities, and the technology has been adapted by many countries as an alternative method for utilizing the by-products as high-quality ingredients.

Haque et al. (1991) conducted two experiments to evaluate the use of ground whole hens in broiler starting diets processed through the dry extruder (Insta-Pro). In both trials, the extruded whole diet improved growth rate and feed conversion in broiler chicks when compared to the unextruded corn-soybean meal control diet. They reported that no microorganisms were found in the extruded feed.

Figure 2 Insta-Pro Model 2500 dry extruder with the optional side feeder for by-products processing.

Carver et al. (1989) coextruded either raw or ensiled shrimp heads or raw or ensiled squid viscera with soybean meal utilizing a dry extruder (Insta-Pro, Model 2000R). The resulting extruded ingredients were coextruded with additional by-products. They reported a processing cost of US $13.35 per ton of the finished product when raw by-products were used. Ensiling the byproduct raised the processing cost of the finished extruded material to US $24.65 per ton. The nutritional analysis of those marine products coextruded with soybean meal had superior levels of nutrients as compared to the solvent-extracted soybean meal alone.

In two experiments, Tadtiyanant et al. (1993) evaluated the nutritional contribution of dead broilers, dead turkeys, raw feathers, or feathers treated with a chemical and enzymatic premix. Those by-products were coextruded with soybean meal using the dry extruder (Insta-Pro), and the resulting ingredients were formulated into complete broiler starting diets and a comparative corn-soybean meal isonitrogenous, isocaloric control diet. Diets containing extruded dead broilers supported higher ($P < 0.05$) body weight than those receiving the corn-soybean meal diet. The treated feathers diet improved the growth rate as compared to the control diet. In a second experiment, deal poultry and treated feathers extruded products resulted in feed efficiency and growth responses comparable to the corn-soybean control diet.

The author (Said, 1996) reviewed the literature regarding the utilization of dry extrusion technology (Insta-Pro) in processing by-products from the poultry industry. He cited the microbiological quality from all reports indicating sterile finished products as they exited the extruder. Literature was cited on the extrusion of hatchery waste and its commercial use and extrusion of feathers, eggshells, and slaughterhouse by-products. The economical feasibility of by-product extrusion was discussed in that report.

In conclusion, the versatility of dry extrusion has been explored. The popularity of its utilization in processing oilseeds, complete feed, by-products, and specialty ingredients reflects its acceptance worldwide as an economical method of processing that can assure high quality finished products with low capital investment and operational costs.

REFERENCES

Aldrich, C. G. and N. R. Merchen. 1995. "Heat treatment of whole soybeans: influence on protein digestion by ruminants." J. Anim. Sci. 73: (Suppl).

Carver, L. A., D. M. Akiyama, and W. G. Dominy. 1989. "Processing of wet shrimp heads and squid viscera with soymeal by a dry extrusion process." In: *Proceedings of the World Congress on Vegetable Protein Utilization in Human Foods and Animal Feedstuffs*. Champaign, IL: A.O.C.S. Press, 1989. pp. 167–170.

Cheva-Isakarum, B. and S. Tangtaweeipat. 1995. "Utilization of full fat soybean in poultry diet II broiler." Asian-Australasian Journal of Animal Science. 8: 89.

Hancock, J. D., R. H. Hines, and T. L. Gugle. 1991. "Extrusion of sorghum, soybean meal, and whole soybeans improves growth performance and nutrient digestibility in finishing pigs." Kansas State University Swine Day 1991. pp. 92–94.

Haque, A. K. M. A., J. J. Lyons, and J. M. Vandepopuliere. 1991. "Extrusion processing of broiler starter diets containing ground whole hens, poultry by-products meal, or ground feathers." Poultry Sci. 70: 234–240.

Kakade, M. L., D. E. Hoffa, and I. E. Liener. 1973. "Contribution of trypsin inhibitor to the deleterious effect of unheated soybeans fed to rats." J. Nutr. 104: 1772.

Kim, J. H., J. D. Hancock, R. H. Hines, and T. L. Gugle. 1994a. "Roasting and extruding affect nutrient utilization from soybeans in 10–20 lbs. pigs." Kansas State University Swine Day 1994. pp. 58–62.

Kim, J. H., H. D. Hancock, R. H. Hines, and M. S. Kang. 1994b. "Roasting and extruding affect ileal digestibility of nutrients from soybeans in growing and finishing pigs." Kansas State University Swine Day 1994. pp. 176–181.

Marty, B. J. and E. R. Chavez. 1993. "Effect of heat processing on digestible energy and other nutrient digestibilities of full-fat soybeans fed to weaner, grower and finisher pigs." Can. J. Anim. Sci. 73: 411.

Mian, M. A. and J. D. Garlich, 1985. "Tolerance of young turkeys to undercooked soybean meal." Fed. Proc. 44: 1524 (abstr.).

Said, N. W. 1996. "Extrusion of alternative ingredients: an environmental and a nutritional solution." J. Appl. Poultry Res. 5: 395–407.

Said, N. W. 1998. "Dry extrusion-mechanical expelling." INFORM. 9(2) (February 1998): 139–144.

Smith, N. E., L. S. Collar, D. L. Bath, W. L. Dunkley, and A. A. Franke. 1980. "Whole cottonseed and extruded soybeans for cows in early lactation." J. Dairy Sci. 63: 153 (abstr.).

Socha, M. I. and M. I. Satter. 1991. "Effect of feeding early lactation multiparous cows heat treated full fat soybeans." M. Sc. Theses. University of Wisconsin, Madison, Wisconsin.

Stillborn, H. L., I. Ndife, B. L. Bowyer, H. M. Hellwig, and P. W. Waldroup. 1987. "The use of full fat soybeans in chicken diets." Poultry Misset International. February 1987: 20–23.

Tadtiyanant C., J. J. Lyons, and J. M. Vandepopuliere. 1993. "Extrusion processing used to convert dead poultry, feathers, eggshells, hatchery waste, and mechanically debond residue into feedstuffs for poultry." Poultry Sci. 72: 1515–1527.

Waldroup, P. W. and K. R. Hazen. 1988. "An evaluation of roasted, extruded and raw unextracted soybeans in the diets of laying hens." Nutrition Report Int. 18: 99.

Wiseman, J. 1984. *Full fat soybeans in diets*. Watt Publishing, Feed International, February 1984.

Woodworth, J. C., M. D. Tokach, R. D. Goodband, J. L. Nelssen, P. R. O'Quinn, and D. A. Knabe. 1998a. "Apparent ileal digestibility of amino acids and digestible and metabolizable energy values for conventional soybean meal or dry extruded-expelled soybean meal for swine." Kansas State University Swine Day 1998. Preliminary progress report.

Woodworth, J. C., M. D. Tokach, J. L. Nelssen, R. D. Goodband, and R. E. Musser. 1998b. "Evaluation of different soybean meal processing techniques on growth performance of pigs." Kansas State University Swine Day 1998. Preliminary progress report.

Zhang, Y. E., C. M. Parsons, K. E. Weingatner, and W. B. Wijeratne. 1993. "Effect of extrusion and expelling on the nutritional quality of conventional and Kunitz trypsin inhibitor-free soybeans." Poultry Sci. 72: 2299.

CHAPTER 4

Interrupted-Flight Expanders-Extruders

MAURICE A. WILLIAMS

A bewildering variety of extrusion devices are used in today's technology. The devices are constructed in many different ways and are given many different names. Squeezing a tube of toothpaste to cause a strand of paste to flow, or extrude, out of the tube gets across the basic idea of extrusion. A hand-driven meat grinder that causes raw meat to extrude as many small strands through a die plate is a more complicated extrusion device. A meat grinder can also have a cutter blade that skims across the die plate to cut the extruded strands into smaller particles or "collets." We are all familiar with these simple devices, and they get across the concept of "extrusion" very well. By use of more sophisticated devices, plastic is "extruded" into tubing and moldings, and precooked wheat is extruded into spaghetti.

Some extruders heat the material as it extrudes and cause the material to "expand" upon extrusion. These special kinds of extruders are used in the food and feed industries to make "expanded" products. Because these heating extruders expand the products, they are often called "expanders."

This chapter is about a special kind of expander that employs an interrupted-flight wormshaft to propel the material through the device. It can be called an "interrupted-flight expander" or, to use a broader name, an "interrupted-flight extruder." Actually, these machines are designated by the trade names of the manufacturers, some of whom use the words "extruder" or "expander" in the trade name. "Interrupted-flight expander" is a good descriptive name to cover all of them, just as automobile is a good descriptive name to cover all of the various brand names used by the auto manufacturers.

An interrupted-flight expander (Figure 1) is mechanically different from other expanders because it was developed from a screw press (Figure 2). Screw presses and expanders are similar in that a revolving wormshaft pushes the material through a cylindrical barrel and out through an opening at the barrel's end. However, a screw press is a more massive and costly machine; it generates more pressure; and it has a slotted-wall barrel that permits oil to flow out from the solids.

Interrupted-flight expanders are used to make pet foods, floating and sinking fish feeds, and full-fat soy. They are also used to prepare oilseeds for solvent extraction and for mechanical crushing. Interrupted-flight expanders are also used to serve as mechanical heat exchangers to dry synthetic rubber. The interrupted-flight expander is an uncomplicated machine, is easily operated, is easily serviced, and can make quality products without a preconditioner.

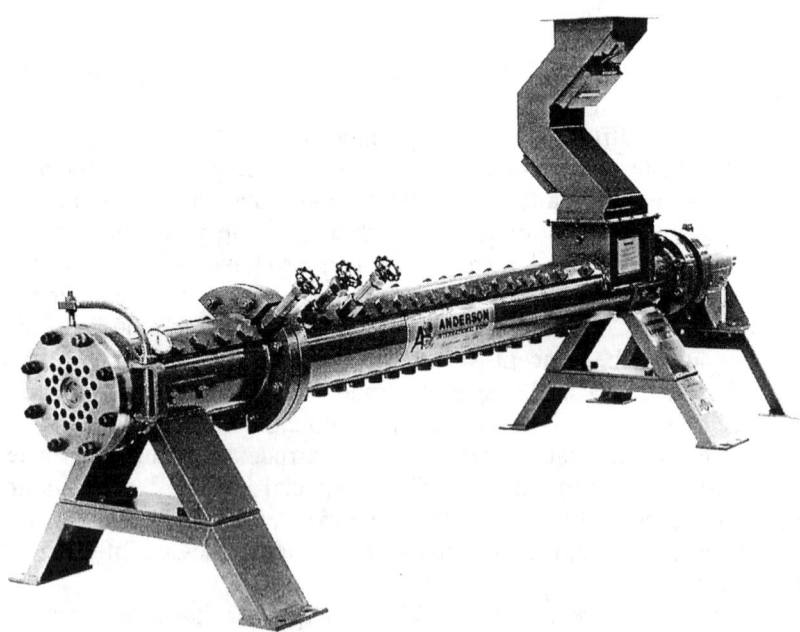

Figure 1 Interrupted-flight expander (courtesy of Anderson International Corp.).

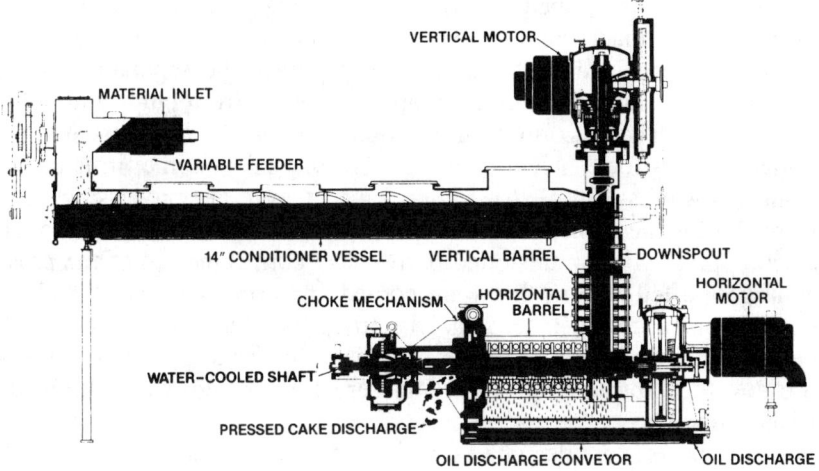

Figure 2 Screw press (courtesy of Anderson International Corp.).

In 1963, Zies and Baer patented the first interrupted-flight expander/extruder borrowing from the technology they used to build screw presses (Zeis, 1963). The first few expanders were used to cook and puff cereal grains for the pet food industry (Williams et al., 1977). Later, these expanders were used to produce dog foods, and then, were used in other industries (Williams, 1993a).

In 1965, Zies adapted the expander to dry synthetic rubber (Zies, 1965). In 1966, Baer, Williams, and Zies patented a process for expanding rice bran to inactivate lipase and bind the bran into sponge-like collets prior to solvent extraction (Baer et al., 1966). In the 1970s, this expander was also used to transform flaked soybean and other oilseeds into porous collets (Williams, 1995b).

BACKGROUND OF INTERRUPTED FLIGHTING

Early experimentation with screw presses showed that reducing the discharge opening (choking) of machines with continuous-wrap screws situated within smooth-walled barrels would not work. When choking was applied, the material, instead of coming under higher pressure, merely spun with the shaft, and the screw press stopped discharging. One hundred years ago, Valerius D. Anderson discovered that if he employed interrupted flighting on the shaft and used stationary projections from the barrel to intermesh with the flighting, the material would not spin with the shaft. The screw press could then develop high pressure when choke was applied. Interrupted flighting was the breakthrough discovery that made continuously discharging screw presses feasible.

The first applications for screw presses were the separation of oil from oilseeds, fat from animal scraps, and water from pulps and polymers. Shaft speed, worm configuration, and preconditioning are important factors for efficient operation. Around 1950, some firms began using screw presses for extrusion-cooking of corn. The processing steps were to grind the corn, preheat and premoisten it and then send it through a screw press with an unslotted barrel. The corn came out of the press as large, puffed chunks that were cooled and crumbled into "kibbled" corn. Kibbled corn replaced the flaked cereals that were mixed with pelleted meat meal to produce a balanced diet for dogs. Later, the entire dog food formulation, meat and cereal combined, were "kibbled" in the same way.

Screw presses are designed for high pressure. These massively built machines operate at slow shaft speed and low capacity (compared to an expander). When the supplier of these screw presses realized there was a market for "kibbled" products, they redesigned their screw press to provide a new machine, less massively built and capable of higher capacity (by faster shaft speed). Anderson's first "Grain Expander" provided a fourfold increase in capacity at one-half of the cost. The barrel of this new machine, instead of being an assembly of drainage bars, spacers, support rings, and heavy clamps, was simplified as a steam-jacketed tube with bolts screwed through the wall to intermesh with the interrupted flighting.

The first start-up showed that steam jackets could not supply enough heat. Feed formulations have a low heat transfer coefficient: approximately 50 BTUs per square foot of contact area per °F temperature differential. The surface area inside an expander barrel is small compared to product flow. A layer of feed sometimes bakes onto the barrel's inner surface and insulates against heat flow. To get around these limitations, Anderson developed a needle valve for injection of live steam directly into the feed formulation within the expander. This brief look at the interrupted-flight's background may help clarify why interrupted-flight expanders differ mechanically from other expanders.

In the late fifties, Anderson experimented with the interrupted-flight expander on many applications and learned that, despite the dependency of a screw press upon preconditioning, the expander did an acceptable job without preconditioning, a strong point favoring this design. The mixing, macerating action of the interruptions thoroughly and quickly blends the injected steam into the solids. Not needing a preconditioner avoids the cost of a preconditioner and allows the plant to handle a dry, free-flowing material in all of the equipment and conveyors going to the expander.

An expander consists of a rapidly rotating wormshaft within a cylindrical barrel. Material enters one end of the barrel and is forced out through a die plate at the discharge end. Figure 3 shows a sectional view of this expander. An interrupted-flight wormshaft rotates within a smooth-walled barrel. Stationary pins are screwed through the barrel's wall and intermesh with the flights. Rapidly revolving worms between stationary pins quickly blend injected water and steam into the solids. Heat of vaporization released by the steam raises the temperature, as does frictional heat from the revolving worms.

The worms compact and work the material, bringing it to a higher pressure as the mixture passes through the barrel. Changing the temperature and moisture conditions inside the barrel influences the degree of cook. Moisture is 15% to 30% by weight. Temperatures can range from 120°F to 320°F. Moisture is influenced by the amount of water and steam injected. Temperature is influenced by the amount of moisture that is injected as steam and by the amount of horsepower generated by the wormshaft. This is influenced by the number and size of die openings through which the product extrudes. Many large dies mean low horsepower consumption. A few small dies mean high horsepower consumption. Moisture level also influences horsepower consumption.

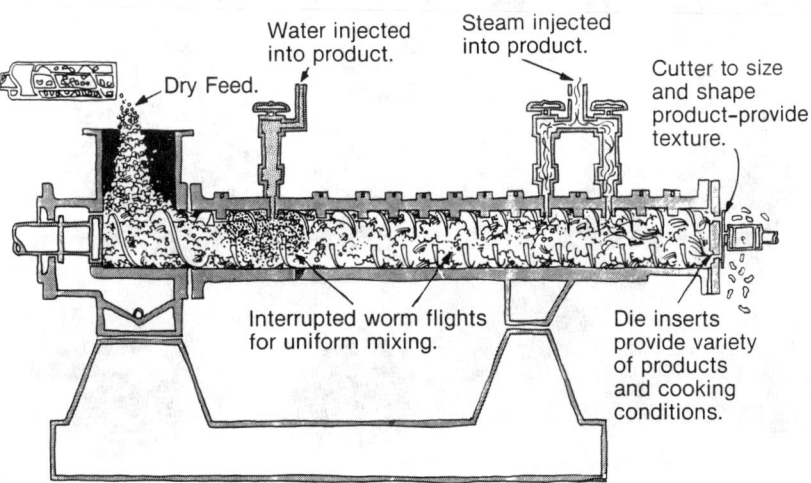

Figure 3 Sectional view of interrupted-flight expander (courtesy of Anderson International Corp.).

The cooking action converts starch into a gel that glues the material into an elastic-like, inflatable mass. When the material leaves the high-pressure interior of the expander, some of the liquid moisture vaporizes, causing the formulation to inflate or expand. The degree of expansion can be influenced by conditions within the expander, but expansion will not happen if there is no inflatable gel in the formulation. This gel is usually starch or gelatinizable protein in animal feeds.

Figure 4 Cutter assembly (courtesy of Anderson International Corp.).

The interrupted-flight expander's advantage is its uncomplicated design and its tolerance for wear. The shaft "floats" within an unjacketed barrel and is powered by a fixed-speed, V-belt drive rather than a gearbox. All the bearings are in a thrust case between the motor and the shaft's feed end. The discharge end "floats" in the material being extruded. The last six worms are coated with Stellite and are surrounded by two replaceable, case-hardened wear sleeves that are inserted in the barrel. Stellite in occasional contact with case-hardened metal insures long life for these wear parts, about the same life as the bearings at the shaft's feed end.

Since the shaft is deep channel (there is a clear space between the hub of the wormshaft and the inner wall of the barrel), and there are no internal pressure plates or steam locks mounted on the shaft, the only restriction to flow is the die plate. When the die plate is opened, the expander can empty itself. A simple clean-out procedure is to send dry feed at high capacity for about 30 seconds to push out any cooked material still remaining when the expander is shut down.

The cutter assembly (Figure 4) is attached to the center of the die plate. The cutter shaft is stationary. The housing and cutter blades revolve. The blades sweep across the surface of each die to cut the product into collets. Collet length is inversely proportional to cutter speed and number of cutter blades. The Collet's cross-sectional area is controlled by die geometry. Length is controlled by cutter speed, and puffing is influenced by how the expander is operated, as will be described below.

MAKING DOG FOOD

Dog foods and other animal feeds are easily made on interrupted-flight expanders. Dogs and cats, being carnivores, cannot digest raw starch. The first requirement in expander-prepared dog foods is to gelatinize the starch. This is easily done. Gelatinized starch forms an elastic, inflatable mass that permits the entire formulation to expand into porous collets upon discharge into atmospheric conditions. Extrusion cooking also inactivates enzymes, antinutritional factors, and bitter flavors. This makes the formulation much more palatable and digestible than when raw.

Some dog foods are extruded through shaped dies and are cleanly cut into shaped particles like the bones shown in Figure 5. The product is dried and bagged "as is" or is coated with fat and sometimes with a gravy powder to add palatability to the product.

Figure 5 Typical dog food (courtesy of Anderson International Corp.).

MAKING FISH FEED

Fish feeds (Figure 6) are made to sink or float by the choice of ingredients in the formulation and by the way the formulation is cooked in the expander (Williams, 1986). Both are important. Neither one can fully counteract the influence of the other. The important variables of cook are moisture and temperature achieved within the expander. These are the two most easily adjusted factors out of five that influence how well a material is cooked. The other three factors are pressure, time, and particle size. An expander operates at pressures considerably higher than atmospheric, usually 200–700 PSI. This high pressure permits cooking in a much shorter time than would an atmospheric cooker. Expanders can operate at different residence times, but, due to high internal pressure, they do not need long residence times. Maximum capacity in a small machine is desired. These two factors dictate that the expander be operated at capacities that result in 20 to 30 seconds residence time.

Feed grind should be 20 mesh, but 40 mesh is better. Any particle larger than 20 mesh won't be fully broken down within the expander. The parameters of pressure, residence time, and grind are important, but they are not mechanically adjusted when the expander is in operation. They are background parameters that are preset before the expander is started.

With moisture and temperature, the higher the moisture, the better the cook; the higher the temperature, the better the cook. A good cook will fully gelatinize starch and convert it into a cohesive, elastic mass that can inflate when the product discharges.

Moisture and temperature levels of 25–35% moisture and 250–280°F thoroughly gelatinize starch. Operating an expander at higher temperature and lower moisture (280–320°F, 15–25% moisture) gelatinizes starch and also converts some of it into dextrine. Higher temperatures are achieved by running at lower moisture and consuming heat generated by friction by the revolving wormshaft. Dextrin influences the properties of the product including its ability to hold water, its stability in water, and its bulk density, and it is an important factor when using expanders to make binders and adhesives from starchy cereal grains.

The moisture level influences bulk density in another way. No matter how much the freshly extruded collets expand upon discharge, if the moisture is high, the collets will collapse after the initial expansion, whereas, if moisture is low, the collets become rigid as soon as they expand, and there is no collapse. Another factor influencing bulk density (and therefore whether a product floats or sinks) is land thickness of the die (Figure 7). Long lands cause the product to be denser and, therefore, more likely to sink.

Figure 6 Typical fish feed (courtesy of Anderson International Corp.).

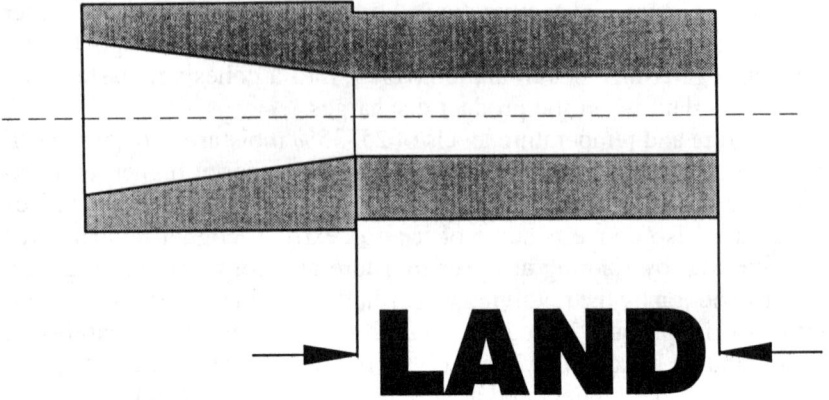

Figure 7 Sectional of die showing land (courtesy of Anderson International Corp.).

MAKING FULL-FAT SOY

Soybean has no starch but has protein that can form a sticky, inflatable matrix that supports expansion into porous collets. Interrupted-flight expanders are used for two different applications on full-fat soybean. One is to transform flaked, full-fat soybean into porous collets for subsequent solvent extraction. The other application is to make full-fat soy for animal feeds.

Using expanders to make full-fat soy is an attractive way for a feed mill to produce its own urease-free, trypsin-inhibitor-free soymeal directly from raw beans. The soybean should be ground. It doesn't have to be dehulled, preheated, or premoistened. Die openings of $1/4$ inch diameter with $1/8$ inch land are sufficient. The trypsin inhibitor level in the extruded full-fat soy is usually 2 to 8 international units per milligram on a fat-free basis. Urease activity can be reduced to undetectable levels, but it is best to see some activity, 0.1, for example, to insure that the soybean has not been overcooked.

Full-fat soy exits the expander as a meal that flashes some of the moisture that had been injected as steam. After flashing, the meal still contains 15 to 20% moisture. Moisture is then reduced in an apron-type dryer/cooler. The air in the dryer is sometimes not preheated; the hot soymeal contains sufficient heat to drive off the moisture, provided it contacts enough low-humidity air to carry the moisture away.

Because expanders cook within 20 seconds, they counteract the activity of troublesome enzymes such as lipase in rice bran (Williams, 1989b) and urease in soybean (Williams, 1991). The short time between enzyme activation and inactivation destroys the enzyme before it has time to cause damage. An expander is even more effective doing this than the horizontal, atmospheric pressure cookers described earlier.

Canola for example, contains enzymes that release phosphorus compounds into the oil. If canola is processed through a stack cooker, the enzymes, which were triggered into action when the canola was flaked, exposed to air, have several minutes to release phosphatides into the oil before the slow-acting stack cooker brings the canola to a high enough temperature to inactivate the enzymes. An expander brings the canola to full temperature in 20 seconds, and the enzymes don't have enough time to release phosphatides. This results in a significant reduction of phosphorus compounds in the finished oil, 48 PPM compared to 350 PPM when using stack cookers. Reduced chlorophyll, free fatty acid, peroxide value, and greenish color are added benefits.

Zang and his coworkers saw a similar situation with phosphorus in soybean oil. They saw that conventionally prepared soybean yielded degummed oil with 184 PPM phosphorus, whereas extrusion-prepared soybean yielded degummed oil with 67 PPM phosphorus (Zang et al., 1994).

In this application, the expander replaces the cooking tray in a stack cooker and cooks the oilseed at 190–221°F. If the oilseed is going to a solvent extractor, it would continue to a prepress or collet expander (same as flakes from a cooker). If the oilseed is going to a full press, then drying to optimum crushing moisture (2–3%) is still required.

EXTRUSION BEFORE SOLVENT EXTRACTION

The first oilseed that was expanded before solvent extraction was rice bran (Baer et al., 1966; Williams and Baer, 1965). Rice bran is a fine powder, making it very difficult to extract. Rice bran also contains the enzyme, lipase, which splits triglycerides into free fatty acids. Lipase can raise the free fatty acid in the bran's oil approximately 3 to 7 percentage points every day (Williams, 1989b; Williams and Baer, 1965). Cooking in an expander inactivates lipase and converts the bran into porous, sponge-like collets that allow rapid percolation of solvent. Expanders also transform flaked soybean and flaked cottonseed into porous collets that handle better in an extractor. Farnsworth et al. described early research with cottonseed (Farnsworth et al., 1986).

Flaked oilseeds extract well if the flakes are of good quality, but if the flakes are thick or if they are crumbly, or if the feed already contained fines, then extraction suffers. Interrupted-flight expanders convert poor quality flakes and fines into easily extracted collets. Even compared to good flakes, expanded collets offer significant advantages because they are larger, heavier, stronger, and more porous than flakes. This permits faster solvent flow (because of larger particles), greater extractor capacity (because the heavier, more porous collets occupy less space), and better drainage (because the collets are strong enough not to crumble as easily as flakes do) (Lusas and Watkins, 1988).

The oilseed reaches 235°F at 10–13% moisture at the die plate under a pressure of 30 to 40 atmospheres. All of the water (natural moisture, injected steam, and liquid water) is compressed into the liquid phase. When the product leaves the high pressure interior of the expander, some of the moisture flashes to reach equilibrium at atmospheric pressure. The flashing inflates the collets with internal pores and surface cracks, giving the collets a sponge-like structure.

Stronger cooks are used for cereal grain gelatinization and cooking of animal feeds, usually 27% internal moisture and 280°F. The material remains within the expander for about 10–30 seconds, depending on rate of input flow. The cook can be changed at any flow rate by changing the moisture-temperature inside the expander.

Some oilseeds are mechanically screw pressed to remove approximately half the oil and are then solvent extracted to remove the rest. Expanders can transform the press cake into porous collets. Steam injected into the expander raises cake moisture 2–4 percentage points. This must be done before the cake has cooled and hardened. Once cool and hard, the denatured protein in the cake can no longer be transformed into a gelatinous, inflatable condition. If oilseeds of high-oil content are prepressed to 15–30% oil, they extrude similarly to flaked soybean (at 18% oil) and flaked cottonseed meats (at 30–33% oil).

SLOTTED-WALL EXPANDERS

Oilseeds containing more than 30–33% oil cannot be extruded through a closed-wall expander because the oil accumulates within the expander and stops steady state operation. In an effort to reduce oil level, extracted meal is sometimes mixed with the fresh oilseed. This reduces the oil level, but the reprocessing of extracted solids increases the load on the expander, the solvent extractor, and the desolventizer. This may force the plant to run at reduced capacity. Williams developed a slotted-wall expander that can make collets from high oil materials by allowing the liberated oil to escape through the slotted-wall drainage cage (Williams, 1990a) (Figure 8).

Field trials with canola (Williams, 1989a) and other oilseeds (Williams, 1990b) showed that this "Hivex™ Expander" could extrude full-fat oilseeds at high oil levels and produce collets at 20–30% oil. Preparation for most oilseeds is to crack or flake the seeds and heat to

Figure 8 HiVex expander (courtesy of Anderson International Corp.).

approximately 60–70°C while reducing the moisture to approximately 8% and being careful not to denature the protein.

EXTRUSION BEFORE CRUSHING

Interrupted-flight expanders are also used to prepare oilseeds for mechanical crushing. Soybean can be passed through a high-shear interrupted-flight extruder equipped with a rotating cone point at the end of the shaft to shear and force the bean to flow between adjustable jaws. This converts the unbroken beans into a frothy meal of almost fluid-like consistency (Williams, 1993b). Fluidization is caused by moisture boiling through the freshly liberated oil. The soybean, at 10% moisture and ambient temperature, for higher capacity, can be cracked before entry into the expander. High shear developed by the expander pulverizes the soybean and ruptures the oil cells. Heat generated by shaft friction raises the temperature to 275°F. This causes the moisture to flash from 10–14% down to 6–7% as the soybean exits into atmospheric conditions.

These high-pressure oilseed expanders are fitted with adjustable jaw chokes in place of die plates (Figure 9) and are operated under low-moisture, high-shear conditions. Under these conditions, they can transform cracked or uncracked oilseeds, with or without preheating, into frothy, semifluid extrudates that flash to 5–7% moisture en route to the

Figure 9 Adjustable-jaw interrupted-flight expander (courtesy of Anderson International Corp.).

screw press. A rotating cone point inserted between two laterally positioned, stationary jaws generates the shear (Williams, 1993b; Williams 1995a). The shear can be increased or decreased by adjusting the proximity of the cone point to the jaws and the opening between the jaws.

An 8-inch-diameter expander with 150 HP drive can process preheated, cracked soybean at 6,000 lbs/hr, producing full-fat soy with low levels of urease and trypsin inhibitor. Also, the hot product can pass into a screw press (after its moisture is allowed to flash to 5–7%). Since most of the oil cells have been ruptured by the expander, a screw press can process the soybean at three times the capacity it would have had with unexpanded soybean.

EXTRUDER DRYING OF SYNTHETIC RUBBER

Another application for interrupted-flight expanders is to serve as mechanical heat exchangers for drying synthetic rubber. This is so different from the previously described applications that some background explanation is needed. Unlike animal feeds and oilseeds and cereal grains, rubber polymers do not need to be cooked or expanded. The only requirement is to remove moisture.

Synthetic rubber is made by polymerization of hydrocarbon gases into long chain polymers. This is done in a solvent media. Once polymerization is complete, the solvent is removed by steam stripping the polymer (and thereby replacing the solvent with water). When the solvent is fully removed, the mixture of polymer and water looks exactly like milk. It is called latex. It closely resembles the natural latex that is the sap of the rubber trees from which natural rubber is made.

The latex is coagulated by adding salt or changing the pH of the mixture, and the polymer forms curds that look like cottage cheese. When the rubber curds (called crumb) are free-drained, they still contain 50–70% moisture. The crumb is pressed in a screw press to reduce the moisture to around 15–20%. The remaining moisture cannot be pressed out. It has to be driven out using heat. Early technology used apron-type dryers (similar to the dryers used to dry extruded animal feeds) to dry the rubber. Zies discovered that an interrupted-flight expander (Zies, 1965) can do a better job of removing the moisture.

The rubber enters the expander at 15–20% moisture. The rubber is unvulcanized and can transform into a viscous semiliquid condition that easily flows through minute openings. Vulcanized rubber (like automobile tires) cannot be melted. Its molecular structure breaks down first.

The unvulcanized rubber absorbs frictional heat from the rotating shaft and from steam jackets surrounding the barrel, and it is then forced to flow through dies (still at 15–20% moisture) that produce either thin ribbons or thin threads, which are cut into small lengths. The extruded hot rubber has maximum surface area and minimum thickness. This allows most of the moisture to flash off as the rubber cools. By the time the rubber reaches room temperature, its moisture has been reduced to around 0.3%.

These rubber-drying extruders (Figure 10) are operated so that just enough heat is transferred into the rubber to cause most of the moisture to evaporate. If too much heat is transferred, the rubber will become hot enough to undergo molecular breakdown. If too little heat is applied, the rubber will still be wet when it reaches room temperature (Dunning and Baer, 1961; Strop and Briggs, 1974).

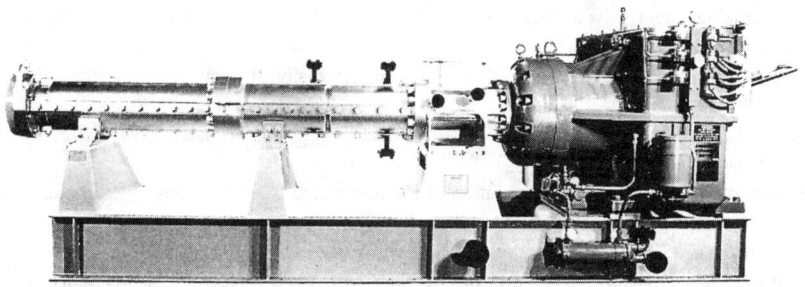

Figure 10 Interrupted-flight extruder-dryer (courtesy of Anderson International Corp.).

SUMMARY

Interrupted-flight expanders evolved from interrupted-flight screw presses, and that's why they differ mechanically from other expanders. The interrupted-flight principle was the breakthrough discovery that allowed a mechanical screw press to operate on a continuous basis, and it revolutionized the oilseed industry, permitting a shift from batch-operated and labor-intensive hydraulic presses to less labor-intensive screw presses that can accept feed and produce oil and de-oiled solids on a continuous basis.

After successful application in screw presses for the oilseed and animal by-products industries for fifty years, this interrupted-flight design was applied to extrusion. The intimate mastication caused by worms revolving between sets of stationary pins, worm after worm all the way through the extruder, causes any injected steam, water, or other liquids to rapidly blend into the solids.

The interrupted-flight expander does not require preconditioning to produce a quality product (but preconditioning increases capacity and reduces wear). Because preconditioning is not required and because the design consisting of a steel tube with bolts screwed through the wall is mechanically simple, the interrupted-flight expander is an inexpensive machine to build.

The interrupted-flight expander has been successfully applied in the animal feeds and human foods industries, the binders and adhesives industry, the oilseed industry, and the rubber industry.

REFERENCES

Baer, S., M. A. Williams, and C. W. Zies. 1966. "Pre-treatment of oleaginous plant materials." U. S. Patent 3,255,220 (to International Basic Economy Corp.).

Dunning, J. W. and S. Baer. 1961. "Dewatering and drying synthetic rubber." *Chemical Engineering Progress.* 57: 53–54.

Farnsworth, J. T., L. A. Johnson, J. P. Wagner, L. R. Watkins, and E. W. Lusas. 1986. "Enhancing direct solvent extraction of oilseeds by extrusion preparation." *Oil Mill Gaz.* 91: 30.

Lusas, E. W. and L. R. Watkins. 1988. "Oilseeds: extrusion for solvent extraction." *J. Am. Oil Chem. Soc.* 65: 1109.

Strop, H. R. and G. O. Briggs. 1974. "Rubber handling: crumb to bale." *Hydrocarbon Processing.* 53: 79–84.

Williams, M. A. 1986. "The preparation of floating and sinking fishfeeds by extrusion." *Infofish Marketing Digest.* 4: 43–44.

Williams, M. A. 1989a. "Expanders for high-oil seeds." *Oil Mill Gaz.* 95: 10.

Williams, M. A. 1989b. "Extrusion of rice bran." In: *The Proceedings of the World Congress on Vegetable Protein Utilization in Human Foods and Animal Feedstuffs*, T. H. Applewhite, Ed., AOCS Press, Champaign, IL, pp. 100–102.

Williams, M. A. 1990a. "Apparatus and method for the continuous extrusion and partial deliquidification of oleaginous materials." U. S. Patent 4,901,635 (to Anderson International Corp.).

Williams, M. A. 1990b. "Using expanders to improve extractability." *INFORM.* 1: 959.

Williams, M. A. 1991. "Extruded starter pig feeds." *Feed Manage.* 42: 20.

Williams, M. A. 1993a. "Preparation of oilseeds to improve extraction of fats." *Extrusion Commun.* 6: 12.

Williams, M. A. 1993b. "Adjustable-jaw expanders." *Feed Manage.* 44: 25.

Williams, M. A. 1995a. "Advancement of processing hardware." *Oil Mill Gaz.* 10: 20.

Williams, M. A. 1995b. "Extrusion preparation for oil extraction." *INFORM.* 6: 289

Williams, M. A. and S. Baer. 1965. "The expansion and extraction of rice bran." *J. Am. Oil Chem. Soc.* 42: 151.

Williams, M. A., R. E. Horn, and R. P. Rugala. 1977. "Extrusion, extrusion, extrusion." *Food Eng.* 49: 99 (Part 1) and 49: 87 (Part 2).

Zang, F., S. S. Koseoglu, and K. C. Rhee. 1994. "Effect of expander process on the phospholipids in soybean oil." *J. Am. Oil Chem. Soc.* 71: 1145.

Zeis, C. W. 1963. "Apparatus for the preparation of feed compounds." U.S. Patent 3,108,530 (to International Basic Economy Corp.).

Zies, C. W. 1965. "Methods for the removal of oil from polymeric materials." U. S. Patent 3,222,797 (to International Basic Economy Corp.).

CHAPTER 5

Twin-Screw Extruders

GORDON R. HUBER

INTRODUCTION

THE process of extrusion has been practiced for well over a century. It involves pushing an extrudate through an opening to produce a predefined shape. The extrusion mechanism can be a simple piston contained within a cylinder, which is capped with a shaping orifice referred to as the die. Material is loaded into the cylinder, the piston moves forward creating pressure at the die, and the material thus emerges in its shaped form from the die. This type of extrusion process is batch in nature because the piston must be retracted periodically to permit refilling of the chamber. Furthermore, the heat buildup in the extrudate is usually limited to the viscous energy dissipation in the die (Hauck and Huber, 1989). This process is considered to be a discontinuous or batch-type extrusion process.

The extrusion process can be made continuous by replacing the piston with a helical screw. Material is fed continuously into an inlet hopper and transported forward by the rotation of the screw. As it reaches the die, the pressure increases to the level required to propel the extrudate through the die orifice. The rotating screw moves material from inlet to discharge as a result of the material slipping on the screw surface. The friction between the material and the screw surface results in heating of the material. These continuous screw extruders can be either single screw or multiscrew in design.

Many different materials are formed and shaped through the process of extrusion. In the polymer industries, thermoplastics, thermosets, and elastomers are extruded. In the food and feed industries, breads, cereals, feed diets, pasta, snacks, confections, meats, starches, and numerous other items are dependent on extrusion. The pharmaceutical and nutraceutical industries use extrusion. And, there are many other industrial applications including, metal, clay, and ceramics.

BACKGROUND

The first commercial uses of extruders appear to have been in the rubber industry in the late 1870s for England and the early 1880s for the United States. These extruders were combinations of ram extruders and screw extruders of short length-to-diameter (L/D) ratios. Around 1935, the basic principles of twin-screw extrusion were conceived and applied to the thermoplastics industry.

The first major commercial application of the single-screw extruder in the food processing industry was the conversion of semolina or flour into pasta. This low-shear, low-temperature-forming process first found commercial production in the 1920s and 1930s and remains a standard production process into the 1990s. Conventional pasta products are processed with an extruder only to the level necessary to bind the moistened mass together and produce a desired shape.

EXTRUSION COOKING

Extrusion cooking has been defined as "the process by which moistened, expansile, starchy, and/or proteinaceous materials are plasticized and cooked in a tube by a combination of moisture, pressure, temperature, and mechanical shear" (Smith, 1976).

In the mid- to late 1940s, the first extrusion-cooked, expanded food products—corn snacks—were commercially produced using single-screw extruders. They were heated sufficiently within the extruder to completely gelatinize the starch. Because of the temperature and moisture conditions that exist during extrusion processing, an exothermic post die expansion of the product takes place. This results in a light density, crisp, textured product.

The early 1950s saw the first application of the single-screw extruder for the production of dry expanded pet food in the United States. This extrusion market has grown into the largest single commercial application of extrusion cooking.

The 1960s brought the first commercial production of dry expanded ready-to-eat breakfast cereals using the single-screw cooking extruder. Traditional breakfast cereal production methods fought hard to compete in limiting the use of extrusion cooking in this product area. Steady improvement in process control, equipment design (including the introduction of twin-screw technology), and better scientific understanding of the extrusion process and raw material behaviors have kept extrusion cooking technology at the forefront of research in this area.

Textured vegetable protein production using single-screw cooking extruder technology became commercially acceptable in the 1970s and is still widely used in numerous applications as meat analogs today.

The 1980s brought rapid commercialization of the production of feeds for aquatic species using extrusion. This currently represents an area of rapid production growth using single-screw cooking extruder technology.

In 1997, the production of extruded snacks in the U.S. was represented by a $932 million sales volume (Wilkes et al., 1998). It is estimated that this figure represents a production volume of 160,000 tons, with a majority being produced using the single-screw extruder.

In 1998, the total commercial retail value of extruded products in the U.S. was $3.62 billion (Bregenzer, 1998). This represents about 3.7 million tons of production volume, again, the majority of which was produced using single-screw cooking extruders.

As we enter a new millenium, these product areas illustrate only a fraction of the almost limitless number of products that can be made using cooking extruders. Continued research and a better understanding of the extrusion process, raw material characteristics and behavior, as well as energy and labor economics will all work together to increase the marketable products provided using the single- and twin-screw cooking extruder systems.

ADVANTAGES OF EXTRUSION

Extrusion cooking offers several advantages over processes it replaces. The most significant advantage is that the process is continuous. Dry powders can be preblended and fed continuously, in a uniform manner, into an extrusion cooker.

The result is less material involved in the process at any given time. Furthermore, the ability to control quality is maximized because poor quality product is recognized immediately. Corrective action is easily taken, and the process is brought into control with a minimum amount of poor quality product produced.

Several processing steps are combined in the extrusion cooker. For example, water, steam and other liquids, and vapors as well as solids can be continuously and uniformly combined. Because several processing steps are combined, the need for extra pieces of complex equipment is eliminated, and the installations are compact compared to other processing means.

The overall utility consumption for extrusion is less than that of alternate processing. This is primarily the result of using lower moisture levels for cooking and shaping a final product. Second, less heat is lost to the surroundings.

Manpower requirements are usually less when using the extrusion cooker versus other processing methods. Large commercial extruders require little attention during operation. Modern process controls are aids in further reducing manpower requirements for production operations.

Extrusion is versatile. It provides precise hardware and process control that permits the use of a wider range of raw materials to produce a given product. As a result, formulations can be altered, and products are often improved.

PAST CONCERNS

In the past 20–25 years, the corotating twin-screw extruder with intermeshing, self-wiping screws has emerged from the food extrusion laboratory to the production floor quite successfully. The use of so-called "twin-screw technology" initially raised the concern in the minds of many that the single-screw cooking extruder represented technology soon to become outdated. Those using the extrusion cooking process were concerned that single-screw extruders may no longer be the most efficient means of producing their products. Many had to decide on the best type of extruder to be added in a new plant expansion. Some processors made the decision to replace their existing single-screw extruders with twin-screw cooking machines.

The questions are still valid. What sets the twin-screw extruder apart from its single-screw counterpart (Hauck, 1988)? And what, if anything, makes it a better processing tool for the production of food and feed products? The answers are not always simple. These questions can be best addressed by discussing some of the equipment and process differences and by examining some of the weaknesses, strengths, and costs of each.

Three points must be considered in evaluating this equipment in any production process:

(1) Shortcomings of the single-screw cooking extruder, which has had over 40 years of success

(2) Improvements offered by the twin-screw extruder
(3) Cost (capital equipment and operation) of these improvements

EXTRUDER CLASSIFICATION

Extruders, single- or twin-screw extrusion mechanisms, tend to be placed in one general category: food or feed extruders. This can be very misleading. Each type of extruder has a very distinct operating principle, function, and application in processing.

Cooking extruders can be classified thermodynamically, by pressure development, or by shear intensity. From a theoretical thermodynamic point of view, extruders have the following properties:

(1) They are autogenous (nearly adiabatic), generating their own heat by conversion of mechanical energy in the flow process.
(2) They are isothermal, meaning that they display a constant temperature.
(3) They are polytrophic, operating between the autogenous and the polytrophic, with part of the energy generated from mechanical dissipation and part from heat transfer.

These classifications become important only when modeling the behavior of the cooking extruder. This modeling is very complex because nearly all cooking extruders operate polytrophically.

Food extruders can also be classified by their method of pressure development—positive displacement or viscous drag. Single-screw, corotating twin-screw, and nonintermeshing cooking extruders are viscous drag extruders. They depend on friction between the screw and barrel surfaces and the extrudate for conveying and pressure development.

Twin-screw extruders can be further subdivided into several classifications (Johnston, 1978):

(1) Corotating intermeshing (Figure 1)
(2) Corotating nonintermeshing
(3) Counterrotating intermeshing (Figure 2)
(4) Counterrotating nonintermeshing
(5) Conical intermeshing

The intensity with which the extrudate is sheared may be the most practical way by which to classify cooking extruders. This permits easy identification and cross-reference of products to process variables and physical parameters of the extruder. Numerous subclassifications also exist (see Chapter 2, "Single-Screw Extruders").

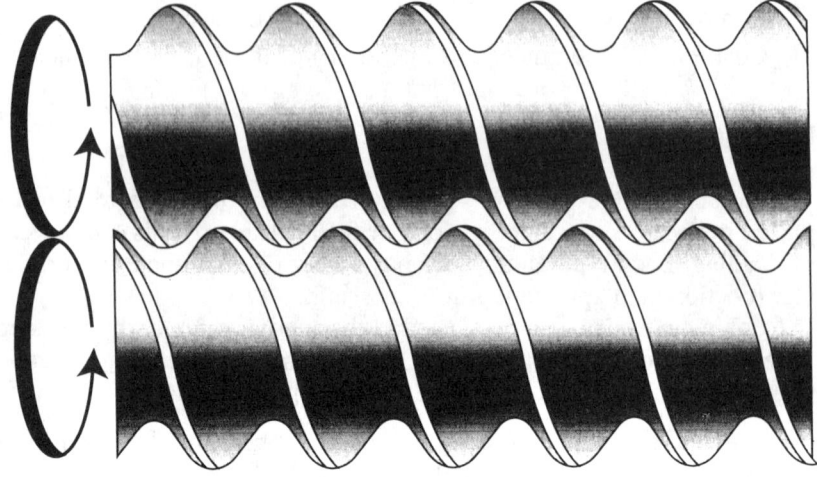

Figure 1 Corotating intermeshing screws.

Figure 2 Counterrotating intermeshing screws.

PROCESS DESCRIPTION

High-temperature/short-time cooking extruders are versatile processing machines. They can use a wide variety of raw materials and formulations to produce products that have been previously mentioned, as well as numerous others. The extruder can be operated by controlling specific process variables in order to produce a wide spectrum of engineered foods. Process conditions and recipes can be changed to alter final product characteristics.

The extrusion cooker, whether twin-screw or single-screw, is made up of several subcomponents common to all extrusion systems (Figure 3). These subcomponents and their functions include the following:

(1) A *holding/live bin* for mixed raw material capable of discharging raw premixed dry ingredients continuously and uniformly
(2) A variable *speed metering/feeding device* to feed the raw, dry mixed ingredients uniformly and in an uninterrupted manner at the desired flow rate
(3) A *preconditioner* or preconditioning cylinder in which liquids and/or steam, and/or other vapors may be uniformly combined with the premetered dry (or pretempered) ingredient mix
(4) An *extruder assembly* for which the configuration of barrel segments, screws, and shearlocks has been preselected to properly feed, knead, and cook the dry or premoistened process material

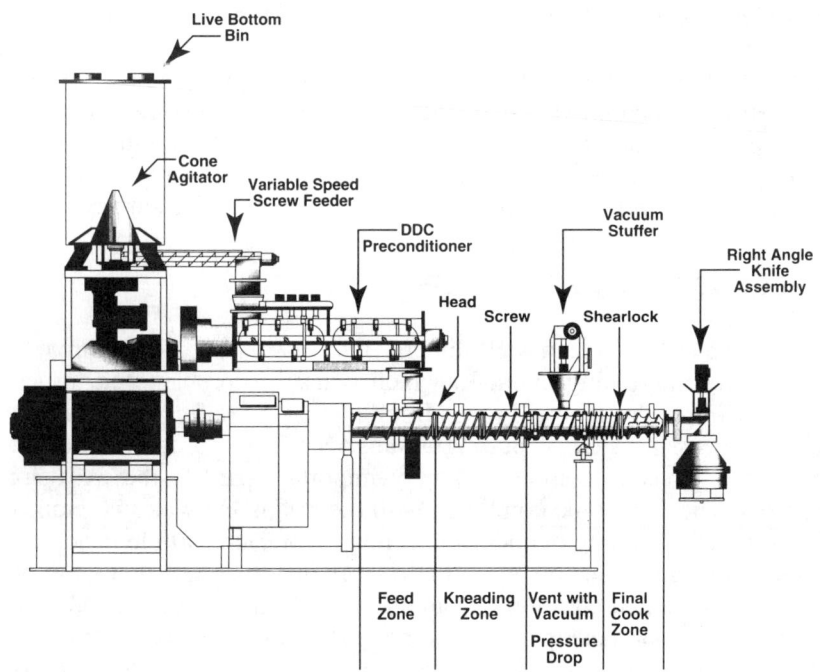

Figure 3 Extrusion cooking system.

(5) A final *die* to restrict the extruder discharge and shape the final product
(6) A *cutting device* to cut the extruded profile to the desired length

DESCRIPTION OF INDIVIDUAL COMPONENTS

HOLDING/LIVE BIN

A holding/live bin provides a buffer of raw material at the inlet so the extruder can operate continuously and without interruption. There must always be a sufficient volume of material in the live bin to permit an orderly shutdown of the system in the event an unscheduled shutdown becomes necessary. This bin must also be equipped with a means of preventing the bin discharge from becoming bridged or blocked off.

METERING/FEEDING SCREW

A variable speed metering/feeding screw must be used to continuously and uniformly discharge material from the bin and feed it to the extruder. Most production size extruders are starve-fed devices. The throughput is governed by the dry feed delivery rate from the feeder screw to the extruder barrel. Loss-in-weight (LIW) feeding systems (sometimes called gravimetic feeding systems) have become a standard for most extrusion processing systems. LIW systems allow the proportioning of dry recipe ingredients and other liquid flows or additives utilized in a final product to be controlled within $\pm 0.5\%$ accuracy.

PRECONDITIONING CYLINDER

The feeder screw can feed directly into the extruder inlet throat or into a preconditioning cylinder, which is used to preblend steam and water with the dry recipe. Preconditioning may be defined as a prerequisite processing step of putting a substance in the proper or desired condition. Preconditioning is a very important part of the extrusion process. The dry recipe combined with the steam and water is retained in the preconditioning cylinder long enough for each particle to achieve temperature and moisture equilibration. Mixing, hydration, cooking, pH modifications, and addition of vapors, flavors, lipids, colors, and meat slurries may all take place in a properly designed preconditioning process. The single most important aspect of the preconditioning system is the added mixing and retention time which is imperative for all reactions, chemical or physical. There are basically two types of continuous preconditioners available, pressurized preconditioners and atmospheric preconditioners.

Pressurized Conditioning

Pressurized conditioning chambers normally provide approximately 1 to 3 minutes residence time at temperatures up to 115°C. Research indicates that pressurized preconditioners have a negative effect on the nutritional quality of food and feed products. Their design and ease of operation are more complex.

Atmospheric Conditioning

Atmospheric conditioning chambers provide from 20 to 240 seconds retention during which time a product formulation is preheated, and the moisture is allowed to penetrate the individual particles.

Preconditioning is known to enhance flavor development and to aid in the final product texture. This has been most evident in the production of cereal or snack products, especially corn- and oat-based products. Oat, for example, has a bitter aftertaste that may be removed with the use of preconditioning. In addition to flavor development, preconditioning reduces extruder component wear, refines cell structure, and permits increased throughput through the extruder.

The method of water addition is also an important part of preconditioning a product formulation. The residence time of the raw material in the conditioning cylinder after water injection with steam injection permits the moisture to penetrate the grain particles equally throughout each particle. This equilibration of moisture provides a more uniformly cooked and textured final product.

The conditioning cylinders typically used in the extrusion cooking processes may be equipped with either a single agitator, dual agitators, or dual agitators with differential speed. Single-agitator mixing cylinders are used in extrusion applications of low to moderate capacity where steam and water are mixed and continuously blended with the dry raw material substrate.

Before purchasing an extrusion system, careful consideration should be given to the effect preconditioning will have on the final product and to how changes or improvements will influence the product's sales potential (see Chapter 6, "Preconditioning").

EXTRUDER ASSEMBLY

The extruder assembly or barrel (Figure 4) is composed of jacketed heads and rotating screws. Jacketing the extruder heads is important to permit modification of the temperature along the length of the barrel. The heads can be heated by steam, hot water, or thermal oil. They can be cooled using water or other cooling media. The extruder bore may be of uniform diameter from inlet to discharge; it can be tapered, decreasing in bore diameter from inlet to discharge, or it can be of uniform diameter with the final segment of the barrel being tapered or decreasing in diameter. The food extruder, whether corotating twin-screw or single-screw, must exert several actions in a very short time under controlled, continuous, steady state operating conditions. These actions include singularly or collectively any or all of the following: heating, cooling, conveying, feeding, compressing, reacting, mixing, "amorphousizing," homogenizing, melting, cooking, texturizing, and shaping.

DIE

The extruder barrel is capped with a die that contains one or more openings through which the extrudate must flow. These openings shape the final product and provide a resistance against which the screw must pump the extrudate. Dies may be designed to be highly restrictive, giving increased barrel fill, residence time, and energy input. They may also be designed for minimal restriction to minimize each of these conditions.

Die design and its effect on functional properties and quality of a

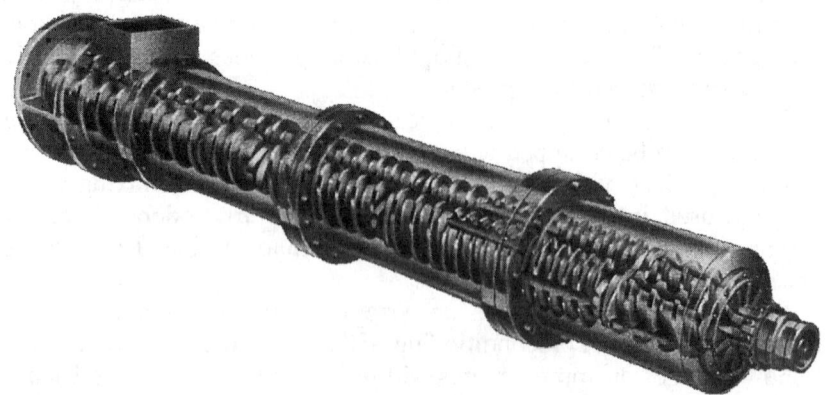

Figure 4 Corotating fully intermeshing twin-screw extruder barrel section.

final product are many times overlooked. Die shear rates may be altered dramatically by changing from a single die with one opening to a triple or quadruple die with multiple openings and flow channels. When changing from a single die configuration to a multiple die configuration, the degree of barrel fill is also increased. Dies with high shear rates cause starch-bearing products to be exposed to increased energy inputs which promotes starch damage resulting in increased water solubility along with other changes to final product characteristics.

Coextrusion dies may take on several different design concepts. The extrusion cooked portion of the product will flow directly through the die parallel to the direction of flow through the extruder barrel. The filling will first be pumped into the die perpendicular to the flow of the outer shell extrudate into the center of the extrudate flow region, then turn 90° to flow with the shell extrudate. The amount of filling pumped into the outer shell may within a small range control the outer product dimensions and the filling-to-shell ratio of the final product. Pulling or stretching of the extrudate may also control the wall thickness of the outer shell, therefore, effecting the filling-to-shell ratio. Careful consideration needs to be given to the filling extrudates that are heat sensitive or contain high lipid content.

Sheeting dies are also commonly used to make 0.8 mm to 1.5 mm thickness sheets up to 3 meters in width. Sheeting dies must be temperature controlled to maintain uniform flow of the extrudate through the die. Stretching of the sheet after exiting the die may be used as a means to control the sheet thickness.

CUTTING DEVICE

Final product length is established by using a knife mounted to run against the die face or a remote cutting device. Alternate post-die devices can be used to establish the final product length and shape. The ability to cut the extrudate into a precise shape definition may determine a product's success or failure. The importance of selecting the proper cutting device should not be overlooked.

EXTRUSION PROCESSING ZONES

The actions and sequence of occurrences in the barrel require the extrusion chambers of the corotating twin-screw and single-screw extruders to be subdivided into processing zones. The feeding zone, the kneading zone, and the final cooking zone (Figure 5) are the most commonly referenced. Other frequently referenced terms include compression, melting, reacting, amorphousizing, and texturizing zones.

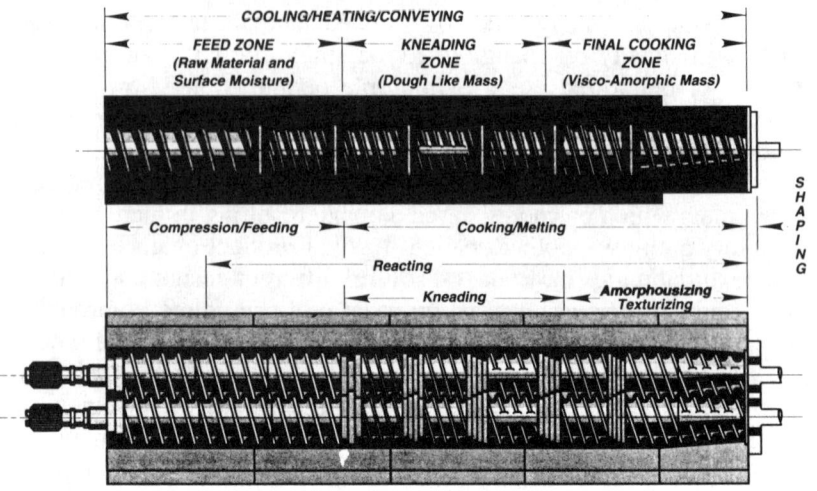

Figure 5 Extruder processing zones.

FEEDING ZONE

This is the area where the low-density discrete particles of raw material are introduced into the barrel inlet. This often preconditioned material is then transported into the interior of the chamber. The flow channel of the screw is typically filled with low-density material in this zone. The density is low due to the air entrapped in the incoming material and due to its granular nature. As the incoming material is conveyed toward the outlet end of the extruder, it is compressed slightly with the air being expelled. Water is typically injected in the feeding zone of the barrel to alter textural and viscosity development and to enhance conductive heat transfer.

KNEADING ZONE

Here the compression started in the feeding zone continues, and the flow channels of the extruder achieve a higher degree of fill as their volume and screw pitch decrease. The extrudate begins to lose some of its granular definition, extrudate density begins to increase, and pressure develops in the barrel. The mechanism of shear begins to play a dominant role because of the barrel fill condition that exists here. Barrel pressure is modest in the early part of the kneading zone. This permits, when desired, the injection of steam at pressures of 5–10 atmosphere. When used, this steam carries thermal energy as well as moisture into the extrudate. The discrete particles of material begin to agglomerate because of the increasing temperature resulting from conduction, direct steam injection, and energy dissipation resulting from friction (Figure 6). Steam brings additional energy into the extrudate, thus increasing capacity and reducing energy costs. Steam is typically injected at 7–9 bar, water is injected at 2 bar. Steam may only be injected into a properly designed extruder barrel and screw configuration. Pressure regions are strategically designed into the configuration to prevent countercurrent flow of vapors back toward the inlet of the extruder and to maximize the utilization of the steam energy. As the extrudate moves through the kneading zone, it begins to form a more integral flowing dough mass, and it will typically reach its maximum compaction. The shear is usually moderate, and the extrudate temperature continues to increase.

FINAL COOKING ZONE

This is the area where amorphousizing and/or texturizing occur. Temperature and pressure typically increase most rapidly in this region where shear rates are highest due to the extruder screw configuration

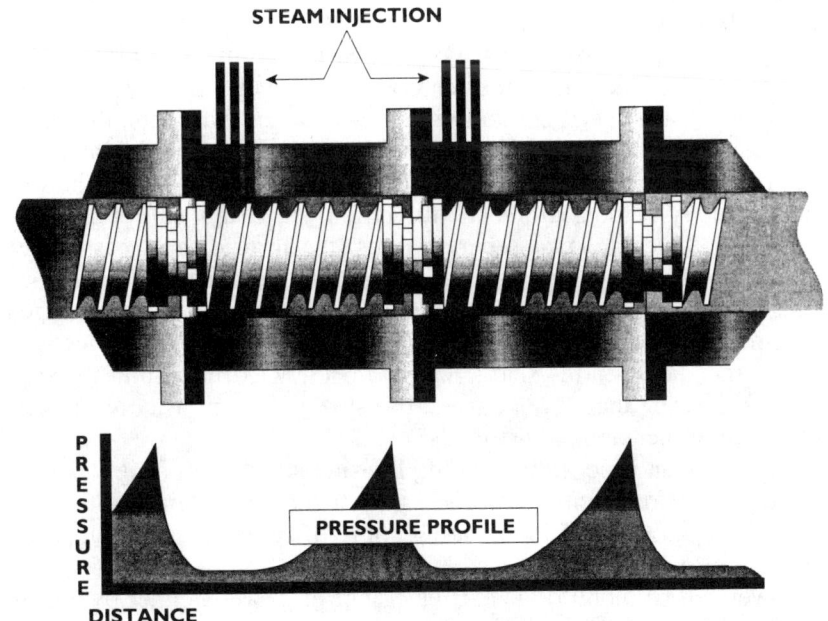

Figure 6 Steam injection.

and maximum compression of the extrudate. The pressure, temperature, and resulting fluid viscosity are such that the extrudate will expel from the extruder die to yield the desired final product texture, density, color, and functional properties. In many twin-screw applications, this zone may be located in the midsection of the extruder. The extrudate then moves to a forming zone where the extrude is discharged from the extruder die in the form of a dense pellet.

TWIN-SCREW DRIVE DESIGN

The screws of the twin-screw extruder can be made of segments approximately as long as the head element. This facilitates manufacture of complicated screw configurations required in the cooking extruder and reduces cost because wearing elements can be replaced individually. Segmented screw elements also provide a means by which the cooking profile of the extruder barrel can be changed to meet modifications in raw materials or final product specifications.

The extruder shaft on a single-screw extruder is supported from or coupled to a bearing housing, which is typically oil lubricated and has either two or three bearings located on the extruder shaft. One bearing, or one for the thrust load of the extruder shaft and the radial load of the shaft, must be placed on one end of the bearing housing, and a radial support bearing must be placed to support the other end. In the single-screw extruder, the physical size of the bearing is limitless, while in the twin-screw extruder, the interrelationship between the two parallel shafts establishes a size limitation. For twin-screw extruders, the bearing assembly is more complicated, and more components such as drive and torque dividing gears are required (Figure 7). To increase the life and simplicity of the gearbox and drivetrain, one manufacturer has separated the drive into two separate units, the thrust-bearing assembly (Figure 8) and the gearbox assembly. This removes the thrust load from being transmitted to the gearbox assembly and allows thrust-bearing replacement to be completed in a matter of a few hours on-site. Because the distance between the shafts on a twin-screw extruder limit the size of thrust bearing that may be used, thrust bearings are usually stacked for handling increased thrust loads.

The lubricant used in the bearing housing should be of high quality and approved for use in food processing. It should contain additives to prevent breakdown of its lubricating ability if contaminated with small amounts of water.

Power can be input to the extruder shaft or gearbox either by a gear train or V-belts. Both methods of power transmission provide adequate speed reduction from the motor speed to the rotation speed required for

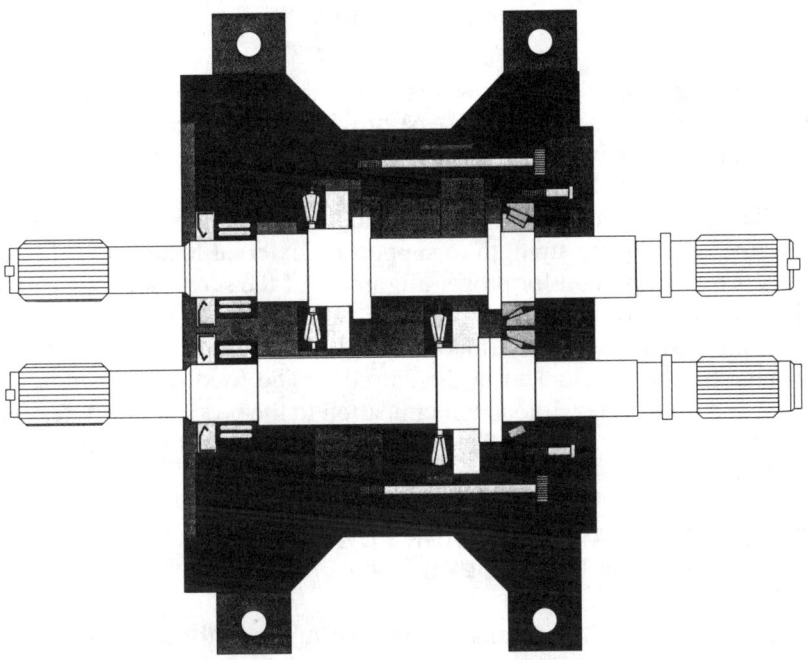

Figure 7 Twin-screw thrust bearing housing.

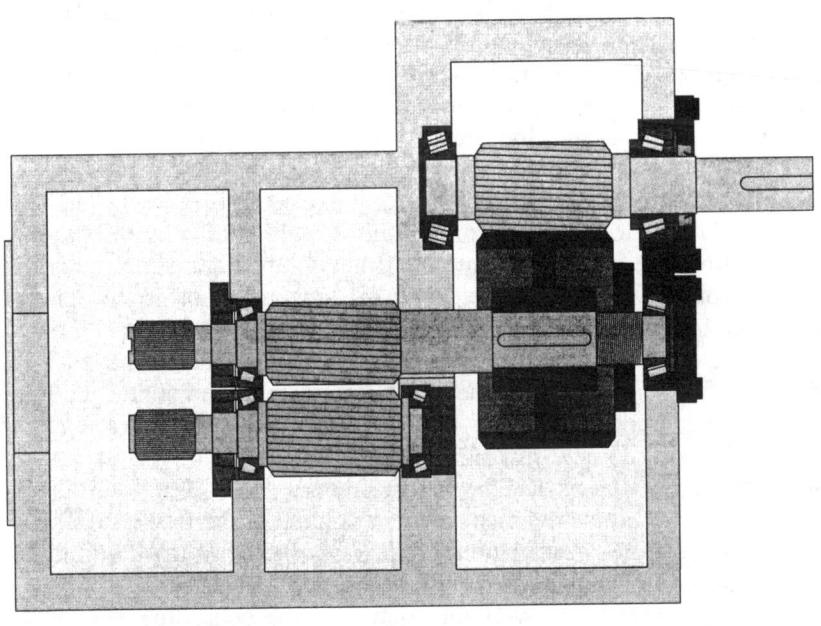

Figure 8 Twin-screw gearbox and drivetrain.

the extruder screws. If the power is input through a gear drive, adequate torque overload protection must be provided to prevent damage to the gears and motor in the event of an overload. V-belts are self-protecting, as they will slip if they become overloaded.

The bearing housing, drive motor, and extruder barrel must all be on a frame of adequate strength to support all external loads applied to it. This is also important for proper alignment of the screw sections within the extruder barrel.

Sanitation is extremely important in the food processing plant, and it must not be overlooked in the extruder. The food extruder must be designed to operate without contamination to the process substrate. Furthermore, it must be possible to clean and inspect all product contact surfaces within the extruder.

SCREW AND BARREL DESIGN FOR SINGLE-SCREW EXTRUDERS

Both the twin-screw corotating intermeshing and the single-screw extrusion mechanisms convey material from inlet to die by having the extrudate slip on the surface of the screw. In the single-screw cooking extruder, the head wall (or barrel wall) can produce insufficient resistance to prevent the extrudate from slipping at the barrel wall, and thus, it will stick on the screw surface and not be conveyed forward. To prevent the extrudate from spinning at the head wall, the inside bore of the single-screw extruder head is often grooved with longitudinal or helical grooves. The helical grooves are configured to propel the extrudate toward the die, and thus, they act also as an extension of the screw helix (see Chapter 2, "Single-Screw Extruders").

Screw design in single-screw cooking extruders is quite varied. The helix of the screw can be of constant pitch and depth from inlet to discharge. Both the screw pitch and the flight depth (Figure 9) usually decrease from inlet to discharge. This is done in an effort to achieve complete barrel fill at the varying extrudate density that is encountered in moving from the inlet hopper to the die. Most extrudates have a bulk density of about 500 g/L in their powdery form as they enter the hopper of the extruder barrel. As the extrudate melts and flows together during the cooking and mixing processes within the barrel, its density increases to as high as 1,800 g/L just inboard of the die. Therefore, it is necessary to reduce the displacement volume of the barrel to achieve a continuous flow. If interruption occurs in the flow mass within the barrel, surging or uneven flow will be observed at the die.

The angle of the screw flight relative to the centerline axis of the screw must also change. In the feeding zone at and immediately after

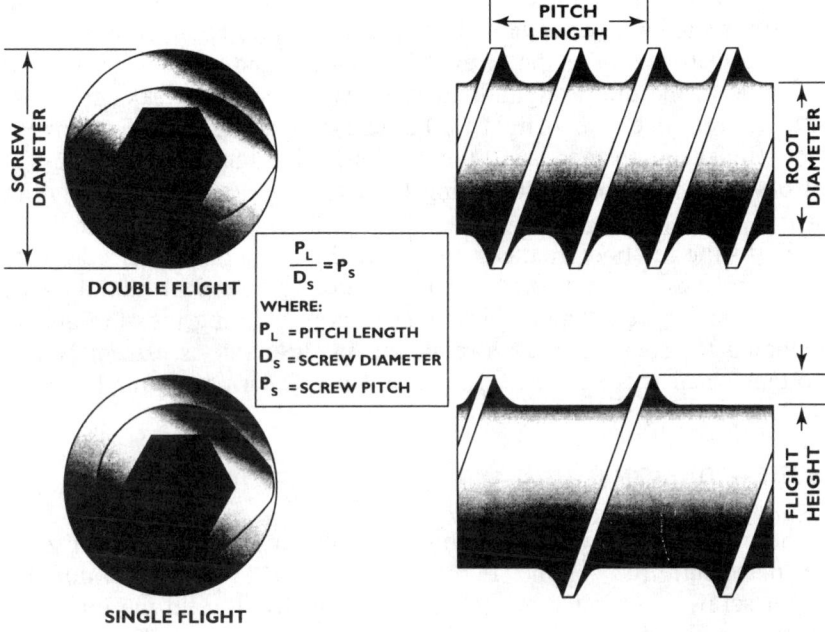

Figure 9 Screw terminology.

the inlet hopper, the screw flight must be at nearly a right angle relative to the screw centerline (see Chapter 2, "Single Screw Extruders"). This promotes the conveyance of relatively low bulk density material. As the extrudate density increases, the angle of the screw flight flattens to increase mixing and decrease conveying efficiency.

SCREW AND BARREL DESIGN FOR TWIN-SCREW EXTRUDERS

In the corotating twin-screw extruder with fully intermeshing screws, one screw flight interacts with the flow channel in the adjoining screw. Therefore, it is not necessary to provide antirotational mechanisms in the barrel walls. If material sticks to the screw surface, the adjacent screw crest will wipe it from the companion screw flank as the two screws intermesh, thus transporting the extrudate forward. This distinctive screw-to-screw interrelationship results in a specific geometric screw flank profile that must be maintained to minimize the buildup of residual material on the screw surfaces.

The barrel wall of the twin-screw extruder is usually smooth but can be constructed with longitudinal or helical grooves. These grooves force the extrudate to slip on the screw flight surface and be transferred from the inlet to the die. Longitudinal grooves increase the leakage of extrudate from one screw flight to the next, thus decreasing the conveying efficiency and increasing the residence time. Spiral grooves assist in building pressure by increasing the pumping action of the rotating screw.

Reducing the bore diameter of the final portion of the extruder barrel—which is referred to as coning or adding a conical section—has been found to be advantageous. This aids in building pressure and in reducing the shear rate. This reduction in shear rate is extremely important when it is necessary to prevent overshearing of fragile extrudate.

SCREW DESIGN

Single-flight screws with pitch ≥1 are often used in the feeding zone for maximum free volume (Figures 10 and 11). Two-flight or double-flight screws with pitch 0.5 to 1.0 are typically used in the kneading section, and either double- or triple-flight screws are used in the final cooking zone. Increasing the number of screw flights increases the screw surface-to-volume ratio, thus increasing the conversion of mechanical energy to heat through friction.

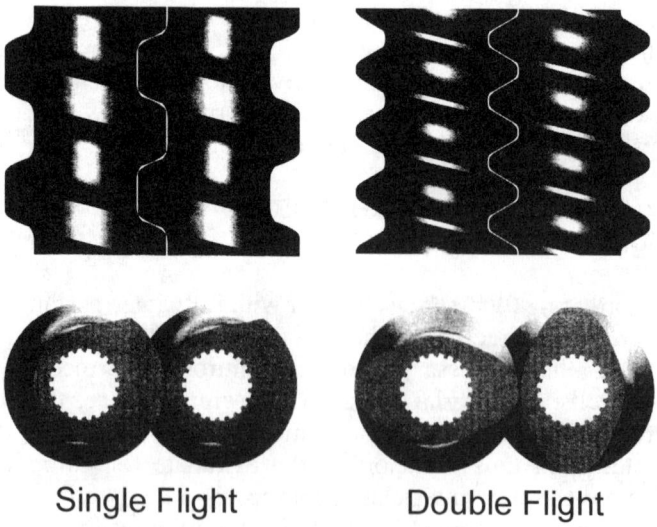

Figure 10 Fully intermeshing corotating screws.

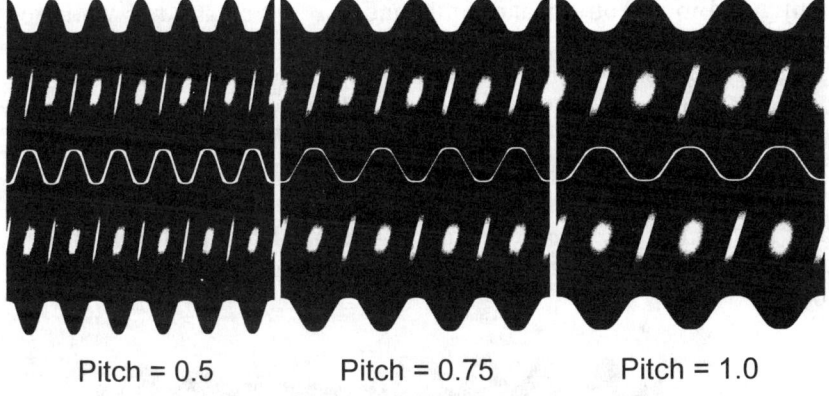

Pitch = 0.5   Pitch = 0.75   Pitch = 1.0

Figure 11 Screw pitch.

Process stability ranks as a top issue facing the user of extrusion systems. Discounting raw material variations and assuming the extrusion cooker is fed a consistent supply of raw material, final product changes can be directly related to uniformity of flow within the barrel and variations in pressure just inboard of the die.

The corotating twin-screw cooking extruder reduces fluctuation of die pressure because of the more positive transport provided by the two intermeshing screws. In contrast to the single-screw extruder, it relies not only on friction, but also on the interaction of one screw flight within the flow channel of the adjacent screw to transport material forward. This reduces the total dependence on frictional forces for transport, and thus, permits the process to operate with a uniform in-barrel flow, resulting in a more uniform die pressure as extrudate viscosity changes.

KNEADING ELEMENTS

The positive pumping characteristic of the twin-screw extruder limits its ability to effectively convert mechanical energy into heat through friction. This obstacle was overcome by the use of reverse flight screw elements and/or by the addition of lobe-shaped shear/kneading elements (Figure 12) to the extruder screw configuration (Figure 4). Either of these elements can reduce the positive conveying feature of the twin-screw extruder and, thus, force the extruder barrel to fill. This allows compression, heating, and shearing of the extrudate as required.

The lobe-shaped shear/kneading (shearlock) elements can be assembled either in a neutral manner (90° increments) (Figure 12) or to assist or repel forward transport of the extrudate. Shearlock elements configured in a neutral manner mix and knead the extrudate while relying on the upstream screw element for transport of the extrudate through the group of shearlocks. Those configured to assist forward transport provide more mixing and kneading than do screw elements, and they provide somewhat less forward transport. When configured to repel forward transport (Figure 12), the lobed elements provide kneading and mixing while tending to pump more extrudate toward the inlet than toward the die. The extruder screw immediately upstream of such a configuration must, therefore, propel the extrudate through this shearlock set.

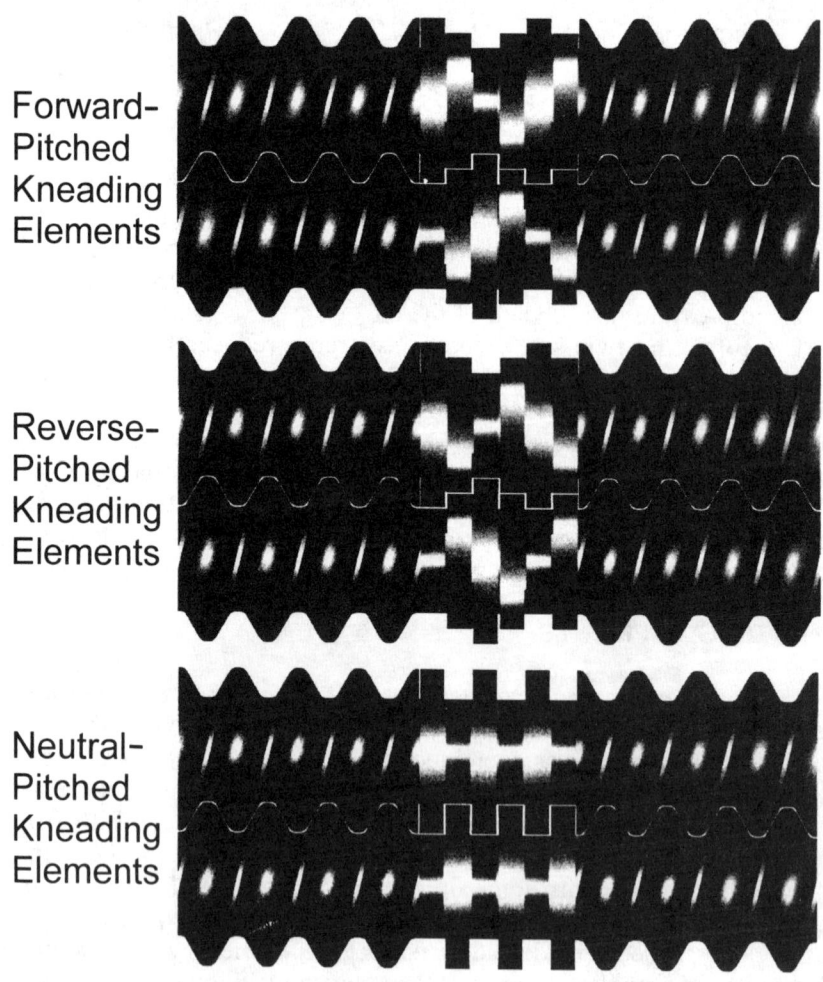

Figure 12 Kneading element orientation.

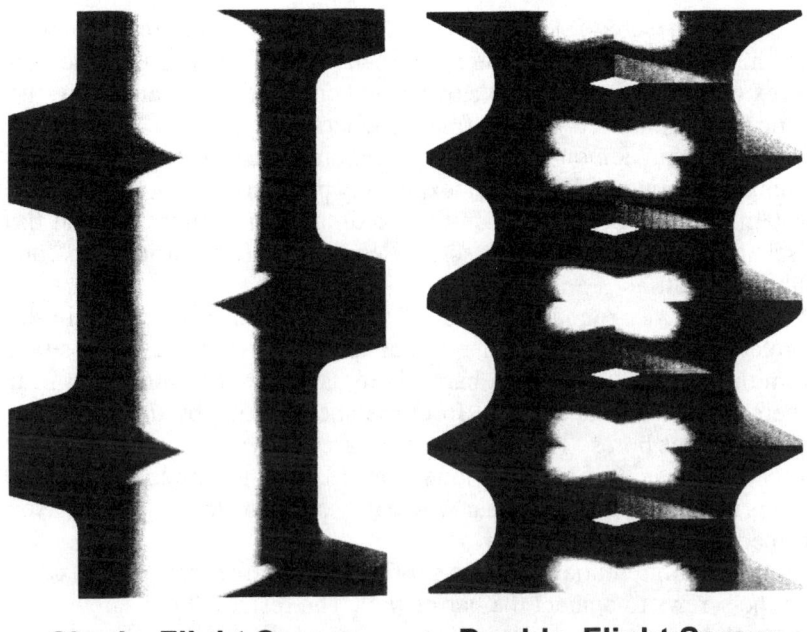

Single-Flight Screw **Double-Flight Screw**

Figure 13 Twin-screw flight profile variations.

Screws may also be used to promote mixing. Various screw patterns such as cut flights or interrupted flights may be used (Figure 13).

REVERSE PITCH

The reverse-flight screw elements provide three negative consequences: high pressures at undesirable points in the extruder barrel, excessive screw and barrel wear (Miller, 1985), and the inability of the barrel to empty. However, there is at least one positive consequence. With reverse-flight screws, the amount of barrel fill within that particular region will remain more constant independent of processing condition fluctuations or changes.

In general, it is desirable for the barrel pressure to increase from atmospheric at the inlet to a maximum pressure just inboard of the die. Detrimental characteristics are often observed in final products if the extruder configuration forces excessive pressure and shear too far upstream of the die.

Excessive pressure and shear can result from reverse-flight screw elements. The pressure will always be highest on the upstream side of a reverse-flight screw element, with the pressure decreasing along its length. This excessive pressure may cause excess power demands on the extruder drive and alter the expansion characteristics (and, thus, the texture and mouthfeel) of the final product.

It has been demonstrated that optimum extruder performance is achieved when making direct expanded products, from starchy and proteinaceous substrates, when the maximum barrel pressure is at the discharge end of the extruder screw (Wenger Manufacturing, Inc., unpublished data).

The pressure around the inside circumference of each screw in the corotating twin-screw extruder is not uniform. At any axial position from the extruder inlet, each barrel bore has a low pressure point near where the screw rotates away from the apex formed by the two intersecting barrel bores. The pressure increases moving circumferentially around the inside of the individual barrel bore until a maximum pressure is reached at the point near where the screw rotates toward the apex formed by the two barrel bores.

This circumferential pressure profile forces screw separation, causing the screws to contact the barrel wall. The result is high screw wear and high barrel wear in a local area on each extruder bore (Figure 14).

When a die is placed adjacent to an extruder barrel with this varying circumferential pressure field, nonuniform die flows result. This yields product of varying length after cutting. Varying product characteristics may also result.

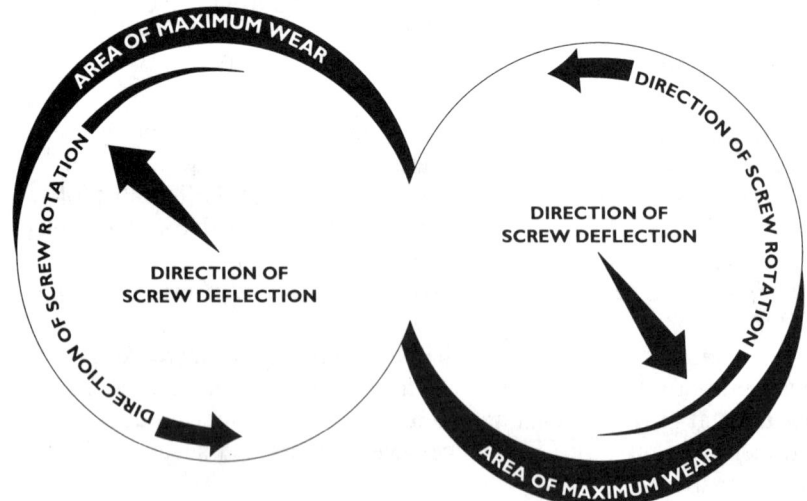

Figure 14 Barrel wear.

Nonuniform die flow can be overcome by using long and/or complicated die spacers. These can cause major quality problems in the finished product because without proper streamlining, extrudate can stagnate in the die spacer, burn, and later break free to contaminate the final product.

CONICAL ELEMENTS

Conical final screw elements added to the discharge of the corotating, fully intermeshing twin-screw cooking extruder eliminate the need for complicated die adapters (Figure 15). These conical elements, which are normally about 1.5 diameters in length, greatly improve performance. Laboratory testing and production history have demonstrated the value these conical final screw elements contribute to the operation of the extruder.

The final conical elements are fully intermeshing at their inlet end. Due to the cone angle relative to the extruder shaft axis, the two complementing conical screws gradually lose intermesh, while they maintain a constant flight depth and uniform flight profile. The conical elements are, thus, not intermeshing at the discharge end. Likewise, the bore of the final head element is configured with two conical bores to

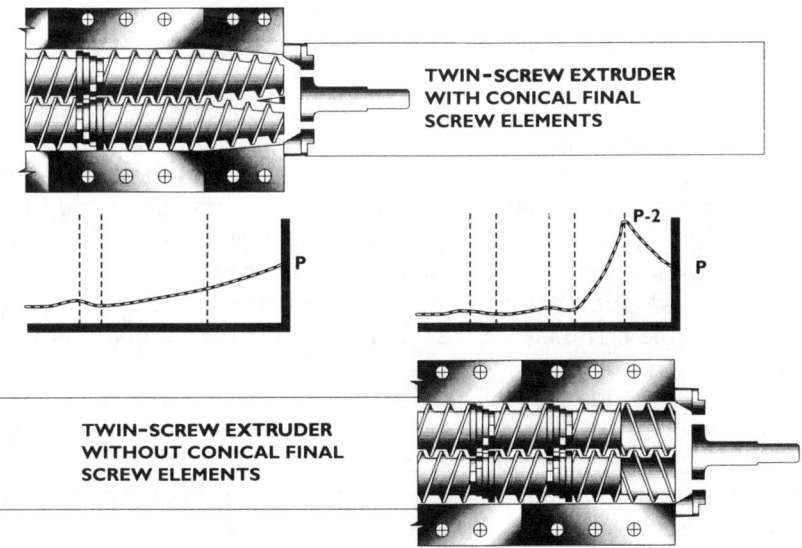

Figure 15 Advantages of conical cone screws.

match the screws. The conical bores in the head reach a point where they are noncommunicative.

Several interesting and rewarding phenomena result from this geometric form. First, the circumferential pressure gradient decreases from a maximum at the inlet end of the conical element to zero near where the screws lose radial communication. There is no circumferential pressure gradient from this point to the end of the screw. Die adapters are, therefore, simplified or even eliminated. Because of the conical configuration of this final section, volume from its inlet to the discharge is significantly reduced. This results in a rapid compression of the extrudate, a rapid increase in barrel pressure, and resistance for the upstream portion of the barrel to push against.

The final conical elements also force a maximum pressure and extrudate density just inboard of the final die. The ability to maintain this high density and the associated pressure in the final section is responsible for the high degree of extruder stability and control.

Pressure at the extruder die is dependent on the design of the extruder or on its configuration, the processing conditions, and the recipe. Pressure at the discharge end of the conical screw has been observed to range from 2 bar (29.5 psig) to 200 bar (3,000 psig). In some process applications, the pressure increase across the conical element may be zero. In applications that demand die pressure of 120–200 bar (1,750–2,000 psig), the pressure increase across the conical element may be 90–120 bar (1,300–1,800 psig).

The condition of the extrudate at this point can be negatively affected by prolonged exposure to excessive mechanical shear, temperature, and pressure. Minimal residence time of the extrudate while being exposed to the most extreme pressure, temperature, and shear environment decreases negative influences on extrudate quality.

Extruder barrel and screw wear have been reduced in corotating twin-screw extruders equipped with conical final screws. By forcing circumferential pressure symmetry in the highest pressure region of the barrel, those extreme screw separating forces that exist in more conventional twin-screw extruders are eliminated. Therefore, contact of the screw with the barrel is reduced, eliminating the typical nonsymmetrical wear (Figure 16).

The conical end sections for the twin-screw extruder offer some unique product application advantages. For example, color may be injected into the conical barrel section at the point the screws are no longer communicating. This allows half of the product exiting the extruder to be a different color than the other half. Using a coextrusion die, unique product may be produced (Figure 17).

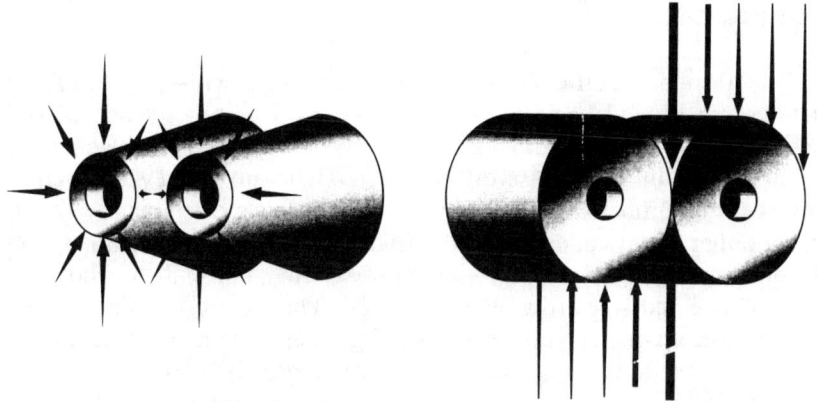

Figure 16 Pressure symmetry.

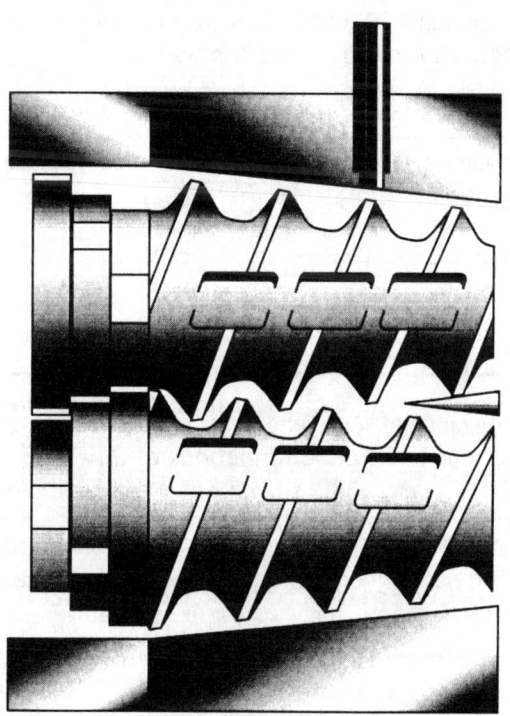

Figure 17 Multicolored products.

DESIGN LIMITS

The design specifications for the corotating twin-screw extruder screws are limited because of the geometric interdependence of one screw relative to the other (Figure 14). The screw pitch can be changed without limit in the twin-screw extruder. There are, however, practical limits. The number of flights can be varied, but most typically this is set at either one, two, or three. Cut flight screws can be used at any position in the barrel. For each twin-screw extruder, the shaft-to-shaft center distance and screw diameter are fixed. Thus, there is a fixed screw root diameter which results in a fixed flight depth. This flight depth will vary from manufacturer to manufacturer and greatly affects the volume of material the screws can convey and the power or torque that may be applied to the screws through the shafts. Shaft design will also play a role in the amount of power that can be transferred to a screw (Figure 18). At a given center distance between the shafts, increasing the screw diameter increases the volume and the flight depth of the screw. Once the number of screw flights and the screw pitch have been fixed, a definite relationship must be maintained to generate the flight profile of the self-wiping, fully intermeshing screw.

Screw Design Parameters for Fully Intermeshing Self-Wiping Extruders.

Design Parameters	Screw Characteristics
Model Parameters • Screw Diameter • Shaft Center Distance	Root Diameter Flight Depth Free Volume
Screw Parameters Screw Pitch Number of Flights	Flight Profile

The center distance to screw diameter ratio is often used to help characterize the volume and torque limitations of a twin-screw extruder. It is also considered when scaling from one extruder to another and during comparisons of extruder manufacturers.

$$\frac{C_L}{D} = \text{Center Distance to Screw Diameter Ratio}$$

where C_L = the center distance between the two shafts or screws, and D = the diameter of one screw.

As this ratio approaches 1.0, the volumetric capacity of the screw approaches zero, and a maximum amount of torque may be applied to the extruder shafts. As this ratio approaches 0.5, the volumetric capacity of

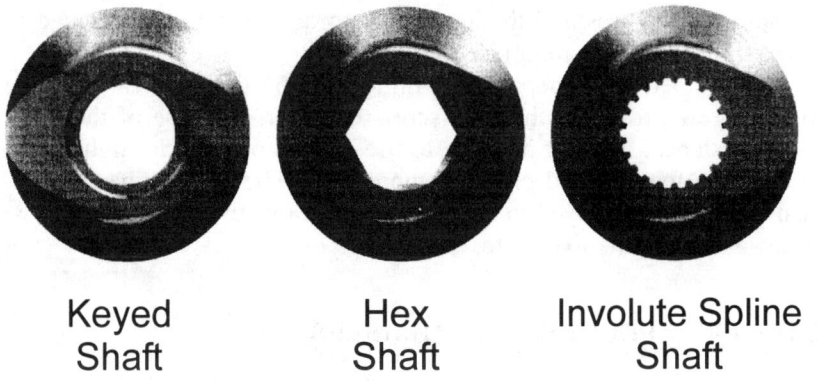

Figure 18 Screw cross sections and shaft types.

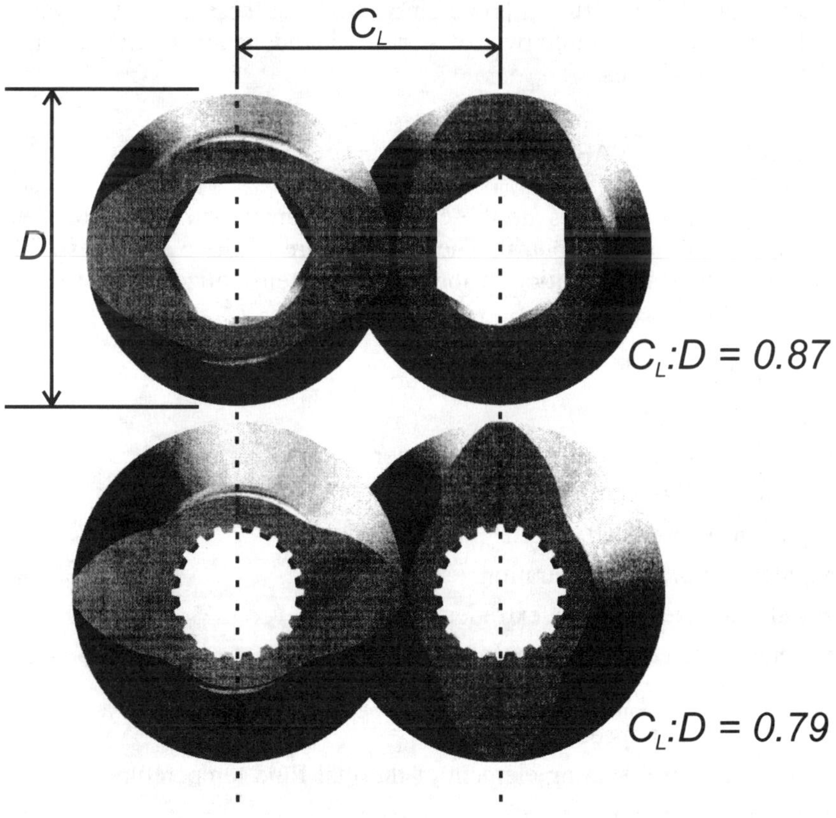

Figure 19 Center distance to diameter ratio.

the screws increases and the amount of torque that may be applied to the extruder shafts is minimized (Figure 19).

This geometric interdependence limits the variations that can be made to the screw profile in the twin-screw extruder. Because of the fixed center distance, it is not possible to use screws of varying flight depth to enhance feeding and create compression. Therefore, other mechanisms must be employed to enable this apparatus to convey, compress, shear, and cook the extrudate.

EXTRUSION PROCESSING VARIABLES

Control of the extrusion processing variables is vital to the success of producing any final product. It is important to understand what processing variables can be controlled directly and which processing variables are simply a result of what is controlled. To understand the interaction of the extrusion processing variables, these processing variables may be divided into two categories: (1) independent variables and dependent variables.

INDEPENDENT VARIABLES

Independent variables are those process parameters that the extruder operator can directly control. The exact nature of these variables will vary with the sophistication of the control system utilized. These variables include the following:

- dry recipe
- dry recipe rate
- water injected into the preconditioner
- steam injected into the preconditioner
- preconditioner shaft speed
- preconditioner configuration
- water injected into the extruder
- steam injected into the extruder
- extruder configuration
- extruder shaft speed
- extruder barrel heating element or thermal fluid temperature
- die configuration

DEPENDENT VARIABLES

Dependent variables are process parameters that change as a result of changing one or more of the independent variables. Dependent variables include the following:

- retention time in the preconditioner
- temperature in the preconditioner
- moisture in the preconditioner
- retention time in the extruder
- temperature in the extruder
- moisture in the extruder
- pressure in the extruder
- mechanical energy input to the extruder

FINAL PRODUCT CHARACTERISTICS

Final product characteristics are measures of final product quality that result from changes made to independent or dependent variables. An initial review of these processing variables and their interactions appears to be rather complex. It may also be overwhelming for an inexperienced extruder operator trying to manipulate these variables to control the final product characteristics. A few of these characteristics include the following:

- moisture—shelf life, stability
- expansion—bulk density, size, shape
- solubility—stickiness, adhesiveness
- absorption—water, fat, milk
- texture—mouthfeel, cell structure
- color—light, dark
- flavor—strong, mild, rancid

CRITICAL PARAMETERS

All final product characteristics are directly influenced by only four critical processing parameters. These critical parameters and the sum of their effects on the raw materials used to make up a recipe determine the characteristics of our desired final product. We interact with these critical parameters through manipulation or changes we make to the independent variables.

These four critical parameters are as follows:

CRITICAL PARAMETER	DESCRIPTION
Moisture	Actual moisture in the process
Mechanical energy input GME = Gross Mechanical Energy SME = Specific Mechanical Energy	$GME = \dfrac{Power}{Mass\ Flow\ Rate} = \dfrac{kWh}{kg}$ $SME = \dfrac{(Power_{Loaded}) - (Power_{Empty})}{Mass\ Flow\ Rate} = \dfrac{kWh}{kg}$
Thermal energy input Expressed in same energy units as mechanical energy Units = kj/kg or kwh/kg	For heating the extruder barrel Thermal fluids Steam heat Electrical heat For direct heating of the extrudate Direct steam injection Other liquid or vapor injection
Retention time $\bar{t}$ = Average retention time m = Amount of extrudate in the process $\dot{m}$ = Mass flow rate	Total time in each part of the process $\bar{t} = \dfrac{m}{\dot{m}}$

If all of the critical parameters are kept constant, consistent duplication of a final product will be experienced. It is important to clarify that this can only happen if the raw materials or ingredients used in the process are consistent. No extrusion system can be expected to make up for variations in raw materials used in an extrusion process.

The following chart may be used as a general guideline for understanding some of the interactions between independent variables and critical parameters. It must be understood that for each of these interactions, one may and will find an exception to the general guideline.

GENERAL INTERACTION CHART

Independent Processing Variables	CRITICAL PARAMETERS			
	Moisture	Mechanical Energy	Thermal Energy	Retention Time
Feed rate	− −	+ +	−	−
Lipid	x	−	x	+
Moisture	+ + +	− − −	+	+ +
Steam energy	+	− −	+ + +	+
Extruder speed	x	+	−	−
Barrel temperature	x	−	+	+
Extruder flow resistance	x	+ +	*	+
Die open area	x	−	−	−

x = no change; + = increase; − = decrease; * = not available

Moisture

Moisture is a critical catalyst in extrusion cooking processes. Moisture in the form of steam, injected into a preconditioning device and into the extruder barrel, brings with it additional energy for cooking. This increases capacity and reduces the requirement for large drive motors. Moisture is necessary for starch gelatinization and protein denaturization. As moisture is increased (within limits), the mechanical energy required for processing decreases. Moisture, in the forms of steam and water, added to a preconditioning device softens the particles of cereal grain, thus, reducing their abrasiveness. This reduces extruder component wear and, in turn, operating costs.

It has been claimed that twin-screw extruders reduce, or even eliminate, the need for adding water or steam to the dry food formula during extrusion cooking. Operating experience has demonstrated that twin-screw and single-screw cooking extruders can process foods under low moisture conditions.

Processing food products at in-barrel moistures below 20% has been proven to be noneconomical as well as nutritionally undesirable. Low-moisture extrusion results in the production of certain undesirable dextrins as a result of increased shear energy inputs. Furthermore, vitamin losses and reduced amino acid availability are greatly accelerated as extrusion moistures are decreased.

Lipids

As lipid levels in a formula exceed 7%, it becomes increasingly difficult to transform mechanical energy into heat necessary for cooking because the lipid acts as a lubricant. The twin-screw extruder mechanism permits the internal lipid level in the formula to be increased to levels beyond 25% while maintaining high levels of mechanical energy conversion. This is made possible by specific screw configurations that are not feasible in single-screw extruders. Steam injection into the extruder barrel is also a contributing factor to cooking formulas high in lipids. Although steam is injected into the barrel of single-screw extruders, and lipid levels of up to 17% can be added to formulations cooked in the single-screw extruder, the twin-screw extruder permits these high lipid level formulas to be processed more consistently. Use of a twin-screw extruder may also result in better lipid binding, and thus, less leaching of lipids from the product during handling and storage.

Screw Speed

The twin- and single-screw extrusion processes are responsive to speed changes in screw. The twin-screw process seems to have a higher responsiveness due to the twin-screw extruder's feeding characteristics. By varying the speed of the screw of the extruder, it is possible to maintain more precise limits on product quality over the wear life of the barrel components. Furthermore, certain variations in raw material characteristics can be compensated for by varying screw speeds.

Because screw speed is such an influential variable in twin-screw extrusion, the twin-screw machine will usually be better for plants making a wide variety of final products at a relatively low volume. Screw speed and die configuration changes can be made along with a formulation adjustment to change the final product. Extruder screw configuration changes are required less frequently in these applications when using twin-screw cooking extruders than when using single-screw machines.

MASS AND ENERGY BALANCE

To understand the interaction between the extrusion process and the critical parameters, a general knowledge of mass and energy balances is important. A mass balance will help determine the total mass flow rate at specific points within the process, calculate moisture content at a specific point in the extrusion system, and calculate flow requirements of input streams to reach a target component content. A mass balance is required for doing an energy balance.

$$\sum \dot{m}_{in} - \sum \dot{m}_{out} = \Delta m_{sys}$$

where: $\sum \dot{m}_{in}$ = Sum of all incoming mass flows
$\sum \dot{m}_{out}$ = Sum of all outgoing mass flows
Δm_{sys} = Accumulation of mass in the system

When the system is operating at steady state, Δm_{sys} = accumulation of mass in the system = 0.

An energy balance will help calculate the temperature of the extrudate at a certain point in the extrusion system, calculate specific energy input into the system, avoid thermodynamically impossible experimentation, and analyze processes on extrusion systems of differing scale or manufacturer.

$$\sum Q_{in} - \sum Q_{out} + \sum \Delta H_{react} = \Delta h_{sys}$$

where: $\sum Q_{in}$ = Sum of all energy flows into the system
$\sum Q_{out}$ = Sum of all energy flow out of the system

$\sum \Delta H_{react}$ = Sum of all energy released by reactions
Δh_{sys} = Change in enthalpy of the system

Again, when the system is operating at steady state, Δh_{sys} = change in enthalpy of the system = 0. For more details on an example of a mass and energy balance, please see the Appendix (Strahm, 1996).

CONCLUSION

This chapter does not define specific products that should be processed using a twin-screw cooking extruder. The intent of this chapter is to provide thought-provoking ideas to help the food processor formulate and answer questions relevant to the purchase of any new extrusion cooking equipment. If a single-screw extruder is currently producing quality products at acceptable profit margins and your market position is favorable, there is no reason to change process equipment. However, if the single-screw extruder is not resulting in a profitable operation, it may be time to evaluate the twin-screw mechanism for that application. The additional cost that may be associated with the equipment and process changes being considered must not be overlooked.

To quote from Janssen's book (Janssen, 1978), "The twin screw extruder is not a panacea." It is an impressive and remarkably versatile and controllable processing tool—perhaps the best our industry has yet seen. Notwithstanding, it will most surely be misapplied to processes for which it is not ideally suited. Not all food manufacturers will require its features and advantages. Those who can are producing premium and novel products that support exceptional profit margins. The applications for single-screw extruders are well established, and many are suited to foods and feeds as formulated today. However, the food formulations of tomorrow may require a different processing approach.

REFERENCES

Bregenzer, B. 1998. "Sales Data." *Petfood Industry*. 40(6): 7.

Hauck, B. W. 1988. "Is the single screw extruder about to become a dinosaur?" *Petfood Industry*. 30(2): 16.

Hauck, B. W. and G. R. Huber. 1989. "Single vs twin screw extrusion." *Cereal Foods World*. 34 (11): 930–934.

Janssen, L. P. B. M. 1978. *Twin Screw Extrusion*. Amsterdam: Elsevier Scientific Publishing Co.

Johnston, G. L. 1978. "Technical and practical processing conditions with single screw/cooking extruders." Paper presented at International Seminar: Cooking and Extruding Techniques. ZDS, Solingen-Grafeath, Germany.

Miller, R. C. 1985. "Low moisture extrusion: Effects of cooking moisture on product characteristics." *J. Fd. Sci.* 50: 249.

Smith, O. B. 1976. "Extrusion cooking." In: *New Protein Foods.* Vol. 2. A. M. Altschul, ed., New York: Academic Press.

Strahm, B. 1996. *Mass and Energy Evaluation in Extrusion Systems.* Applied Extrusion Workshop, University of Nebraska, Lincoln, NE.

Wenger Manufacturing, Inc., unpublished data.

Wilkes, A. P., A. Walter, and D. Hodgen. 1998. "State of the snack food industry." *Snack World.* 55(6): 6.

CHAPTER 6

Preconditioning

BRADLEY S. STRAHM

COOKING extruders were first introduced to the food and feed processing industry in the late 1950s. Since that time, these systems have grown in popularity, efficiency, and flexibility. From the beginning, preconditioning with steam and water has been an important part of the extrusion process. It continues to be so today regardless of whether the preconditioner is associated with a single-screw or twin-screw extruder.

There are many applications of extrusion cooked food products where preconditioning plays a key role in the overall extrusion process. These products include direct expanded and flaked breakfast cereals, precooked pasta, textured vegetable protein, meat analogs, extruded bread crumbs, continuous dough mixing, and third-generation snacks.

Preconditioning prior to extrusion may not be beneficial to all extrusion processes. In general, preconditioning will enhance any extrusion process that would benefit from higher moisture and longer retention time.

The preconditioning process is really a simple one. Raw material particles are held in a warm, moist, mixing environment for a given time and then are discharged into the extruder. Good mixing is required to bring the surface of the raw material particles into contact with the added steam and water. Long retention time is required to allow the processes of diffusion and heat transfer to transport the moisture and energy from the surface to the interior of the particle. This results in raw material particles being plasticized by the steam and water in the preconditioner environment. In practice, the objective is to completely plasticize the raw material particles in order to eliminate any dry core as illustrated in Figure 1.

The principles of polymer science can be used to better understand the effects and benefits of proper preconditioning. It is well publicized that the addition of water and heat to the types of biological raw materials used in the extrusion process will plasticize or soften the materials. The transition that occurs during this process is referred to as the glass transition. The temperature around which this takes place is called the glass transition temperature, T_g. Further heat and moisture addition beyond this temperature will yield to a transformation that will render the material flowable. This transition is called the melt transition and occurs around the melt transition temperature, T_m. These temperatures can be measured by a number of means, but current information indicates that the capillary rheometer method described by Zhang et al.

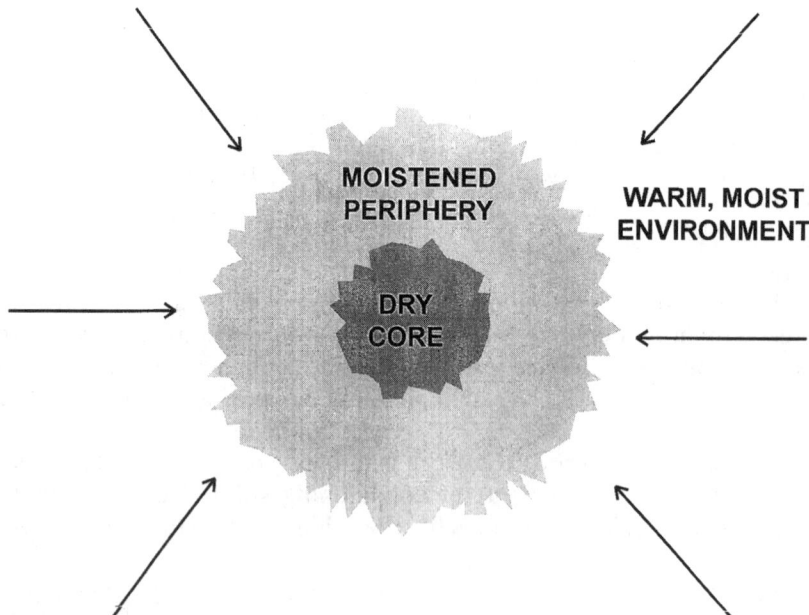

Figure 1 The objective of preconditioning is to eliminate the unplasticized core in the raw material particles.

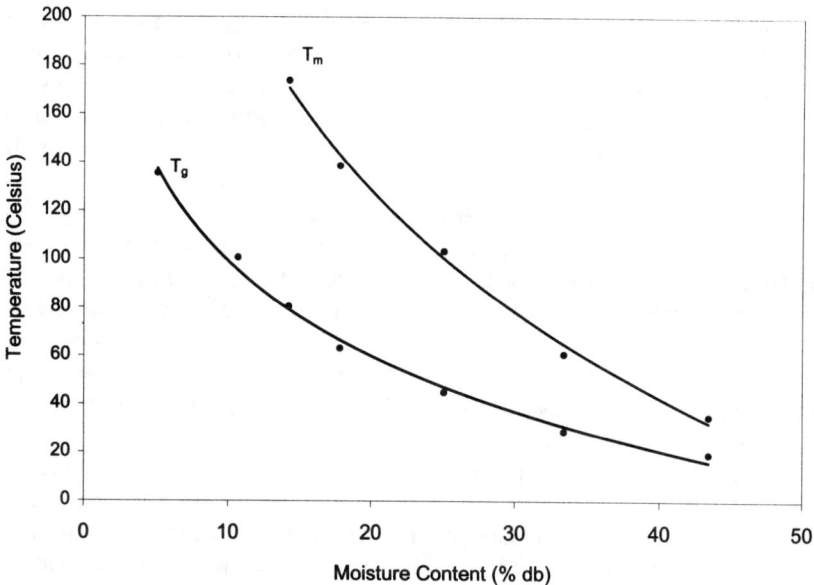

Figure 2 Glass and melt transition curves for cornmeal as measured by capillary rheometer.

(1998) is best for the complex biological materials used in the food extrusion process.

The dependence of glass and melt transition temperatures for cornmeal, a common ingredient in extruded food products, is shown in Figure 2. In a typical extrusion process, the raw materials being added to the preconditioner would be at about 12% (dry basis) moisture and 25°C. From Figure 2, you can see that these conditions clearly indicate glassy, difficult to deform particles. The material being discharged from the preconditioner could typically be about 25% (dry basis) moisture and 80°C. Also from Figure 2, you can see that these materials lie in the rubbery zone between the glass and melt transition lines. Therefore, as the glass transition temperature is lowered by the addition of water and the material is heated to above its glass transition temperature, the raw material particles move from a glassy state to a rubbery state. These particles are then soft, deformable, and easily transformed by the extruder into a final product (Strahm, 1998).

BENEFITS OF PRECONDITIONING

The benefits of adding preconditioning to the extrusion process have been recognized to be fourfold. First, in the area of machine life, preconditioning will increase the life of wear components in the extruder barrel by several times. Second, in the area of extruder capacity, preconditioning has proven to increase the throughput of the extrusion system. Third, in the area of product quality, preconditioning will assist in altering product textures and functionality. Finally, adding preconditioning to the extrusion process enhances product flavor.

Unpreconditioned raw materials are generally crystalline or glassy amorphous materials. These materials are very abrasive until they are plasticized by heat and moisture within the extruder barrel. As already mentioned, preconditioning prior to extrusion will plasticize these materials with heat and moisture by the addition of water and steam prior to their entry into the extruder barrel. This reduces their abrasiveness and results in a longer useful life for the extruder barrel and screw components.

Extruder capacity can be limited by many things, including energy input capabilities, retention time, and volumetric conveying capacity. While preconditioning usually cannot overcome the extruder's limitations in volumetric conveying capacity, it can significantly contribute to energy input and retention time. Retention time in the extruder barrel can vary from as little as 5 seconds to as much as 2 minutes, depending on the extruder configuration. In contrast, average retention time in the preconditioner can be as long as 5 minutes. For some high-moisture processes, the energy added by steam in the preconditioner can be as much as 60% of the total energy required by the process. This reduces the mechanical energy required from the extruder and allows the rate to be increased without adding a more powerful drive motor to the extruder.

Product quality is often related to the physical form of the starch in the product. Starch transformation during the extrusion process is depicted in Figure 3. Products that contain significant amounts of raw, crystalline starch will be raw tasting. Products containing starch that is

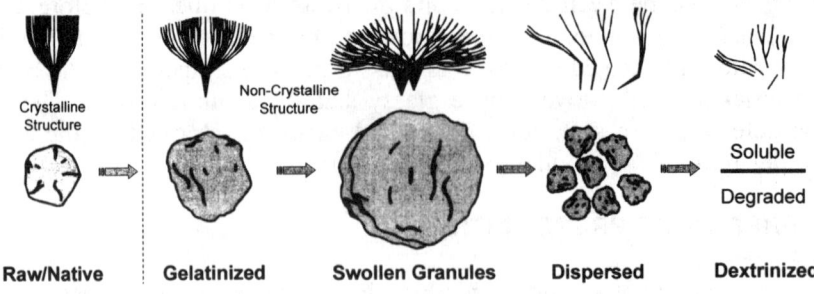

Figure 3 Depiction of the starch cooking and transformation process.

overprocessed, such that they contain dextrinized starch as shown in Figure 3, will have objectionable textural characteristics. For example, in a product that is consumed dry such as an extruded corn snack, a product containing dextrinized starch will tend to stick to the teeth and pack in the molars. In a product that is consumed either rehydrated or together with water in some form such as precooked pasta or breakfast cereals, a product containing dextrinized starch will have a slick or slimy surface texture.

In products that are extruded at in-barrel moisture levels of less than 18% moisture, preconditioning will probably not help to prevent dextrinization, although it can be controlled by other means. In products that are extruded at in-barrel moisture levels above 18%, preconditioning can help to avoid dextrinization. This is accomplished in two ways. First, as already mentioned, it softens the raw material particles, making the starch less susceptible to dextrinization. Second, by adding energy in the form of steam in the preconditioner, less mechanical energy is needed later to accomplish the cooking of the product. Mechanical energy provided from the extruder screws is the force that causes starch dextrinization in the extruder barrel.

Quite often, the extrusion process is used to make a product that is similar to one made by a longer, batch-cooking process. An example of this is breakfast cereals, some of which are traditionally made in cooking processes that exceed 1 hour or more. During the long retention time batch-cooking process, many "cooked grain" flavors are developed as a result of complex reactions among the individual grain constituents. While preconditioning with extrusion still does not have as long a retention time as the batch process, the additional retention time it provides to the extrusion process assists in developing these desirable flavors.

PRECONDITIONING HARDWARE

Preconditioners are mounted in the extrusion process between the feeding device and the extruder barrel. The feeding device provides a continuous flow of dry raw materials to the preconditioner. At the discharge of the preconditioner, a diverter allows the operator to select to either bypass the preconditioned materials to a waste bin or direct them into the extruder. This allows the extruder operator to start the preconditioner first, then divert the preconditioned material to the extruder when its conditions are optimized for extrusion.

Over the history of extrusion in the food and feed industry, two broad categories of preconditioners have been applied. Atmospheric preconditioners operate at prevailing atmospheric pressure and are, therefore, thermodynamically limited to a 100°C discharge temperature. Pressurized preconditioners are sometimes used, but preconditioning at these elevated temperatures and pressures has been associated with increased losses in essential amino acids such as lysine (de Muelenaere and Buzzard, 1969). In addition, pressurized preconditioners are more mechanically complex and, therefore, cost more to purchase and maintain. For these reasons, atmospheric preconditioners are used, almost exclusively, in these industries.

The first extruders introduced into this industry in the late 1950s had single-shaft preconditioners similar to that shown in Figure 4. These preconditioners ran at high speeds to obtain good mixing but had retention times of 30 seconds or less and were for this reason relatively ineffective by today's standards. In the late 1960s, preconditioning technology was updated by the introduction of the double-shaft conditioner like the one shown in Figure 5. These systems had longer retention times of up to 2 minutes and represented a leap forward in technology. The latest, patented, preconditioner design was introduced in about 1986 (Hauck, 1988). This preconditioner, as shown in Figure 6, has a double-shaft, differential diameter, and differential speed design. The design of this system provides average retention times of 2 to 4 minutes. This preconditioner design provides longer retention time in the large, slower rotating chamber while at the same time gives good mixing in the small, fast-turning chamber. The preconditioners shown in Figure 6 range from 2 ft^3 to 108 ft^3 in free volume and are sized for capacities ranging from 300 to 18,000 kg/hr.

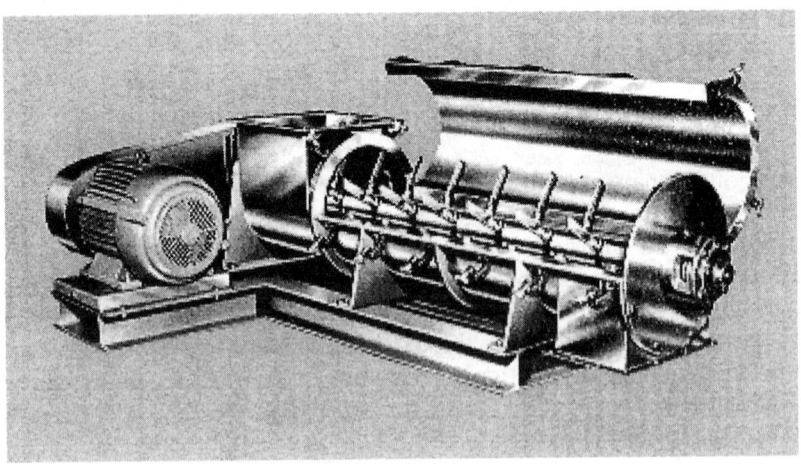

Figure 4 Single-shaft preconditioner.

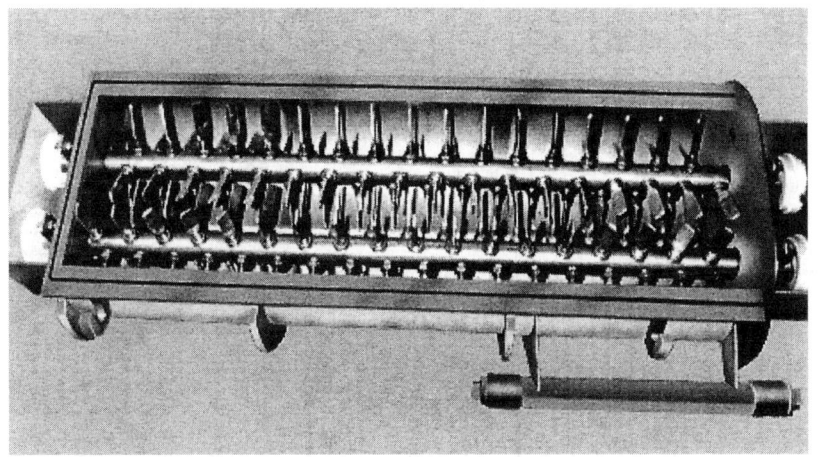

Figure 5 Double-shaft preconditioner.

As shown in Figure 7, dry, saturated steam at a pressure of less than 200 kPa is usually added to the preconditioner from the bottom of the preconditioning chamber and percolates up through the bed of material in the preconditioner. Water is usually added from the top of the preconditioner. The water can be heated to up to 80 to 90°C before being added to the preconditioner. The process of diffusion, which transports the water to the interior of the raw material particles, occurs more rapidly at higher temperatures. Adding water through spray nozzles helps to distribute the water to the raw material particles. This reduces the amount of mixing required in the preconditioner.

Some food products may require the addition of oil or melted fat into the extrusion process. If these materials are added into the preconditioner, they should be added as close to the preconditioner discharge as possible. If they are added before or at the same time as the water, it is possible that the oil may coat the raw material particles and act as a

Figure 6 DDC preconditioners.

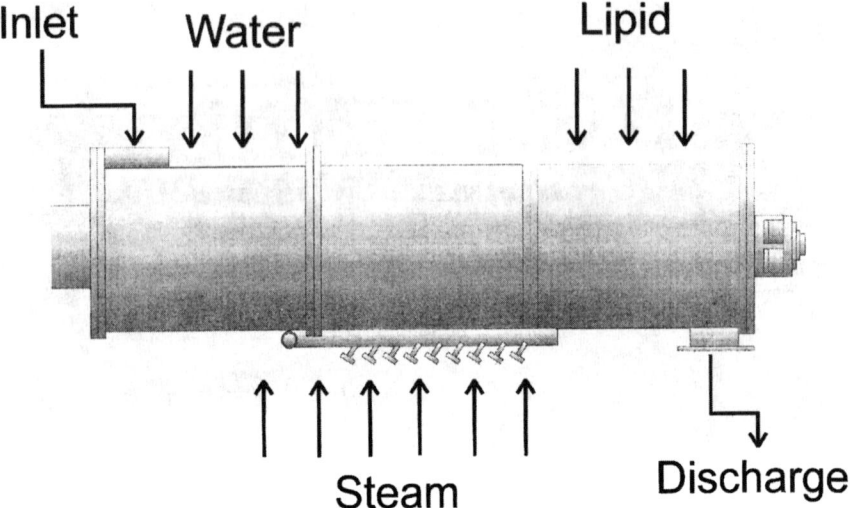

Figure 7 Introducing materials to the preconditioner.

barrier to moisture transfer to the surface of the raw material particles. For products with very high added fat levels, it may become necessary to add the oil closer to the preconditioner inlet in order to have sufficient mixing of the oil with the dry raw material particles. Like water, oil and other liquids can also be added through spray nozzles to reduce the amount of mixing needed in the preconditioner.

PRECONDITIONER OPERATIONS

There are a number of operational variables over which the extruder operator has control when operating a preconditioner. Typical on-line changes include dry recipe rate, steam addition rate, and water addition rate. Other changes that may require off-line modification of the hardware setup include beater configuration and shaft speed. It is important to understand how changes in these parameters impact the effectiveness of the preconditioning process.

As already mentioned, in order for the preconditioning process to be effective, it should hold the raw material particles in a warm, moist environment where there is sufficient mixing to contact the particles with steam and water. In addition, it should provide for sufficient retention time to allow the moisture to completely penetrate the raw material particles.

The amount of total moisture that can be added to the preconditioner is limited by the characteristics of the raw materials. In order for the preconditioning process to be carried out, the raw materials must remain flowable throughout the preconditioning process. Materials that become very sticky when water is added, such as pregelatinized starches, cannot be preconditioned at a moisture content greater than approximately 18% moisture. Other materials can absorb up to 30% or even more moisture and still remain flowable.

Atmospheric preconditioners are theoretically limited to a maximum discharge temperature of 100°C, the boiling point of water at atmospheric pressure. In practice, the actual maximum discharge temperature seems to be in the neighborhood of 95°C. Because of its high specific heat, the amount of water being added has a large influence on the amount of steam that can be added in the preconditioning process. As shown in Figure 8, as more water is added to the preconditioner, a greater heat sink is available to absorb energy from the steam, and therefore, a greater quantity of steam can be added. However, unless the steam is condensed in the preconditioning process, very little heating benefit can be obtained from it.

It is very difficult to measure mixing in any type of continuous mixing device such as the preconditioner (Levine, 1995). It does, however, seem readily apparent that high mixing speeds at longer times will provide better mixing.

It is easier to measure retention time in terms of average retention time and retention time distribution. Average retention time (ART) is a general measure of the overall magnitude of the retention time. It does

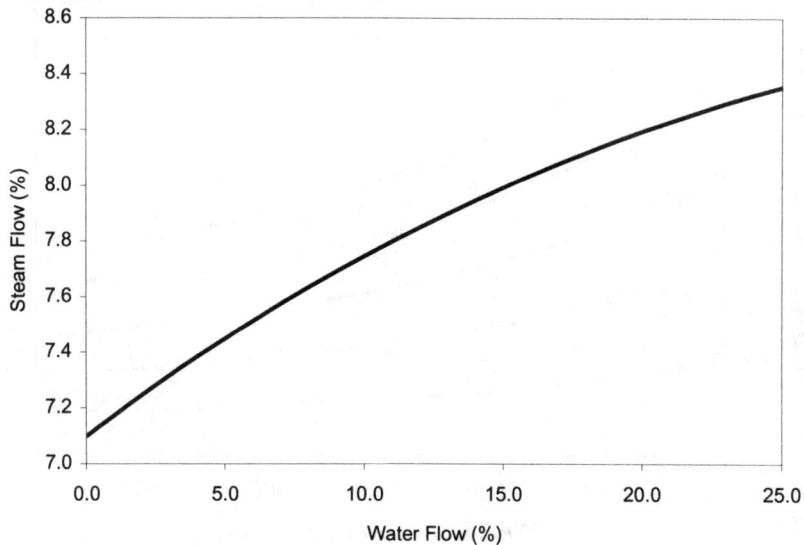

Figure 8 Effect of added water on steam required to reach maximum temperature.

not give any indication of the uniformity of treatment. Uniformity of treatment time from particle to particle is indicated by retention time distribution (RTD). Of these two measures of retention time, average retention time is the more important.

Average retention time is primarily determined by the degree of fill in the preconditioner and the overall throughput. Preconditioners generally run with anywhere from 10% to 50% of their free volume being occupied by the material being preconditioned. The most modern preconditioners are usually designed to operate at the 40 to 50% fill level.

Figure 9 shows the effect of changing throughput and degree of fill on the average retention time in a preconditioner. The degree of fill within a preconditioner is primarily determined by the beater configuration. Configuring more beaters in a reverse conveying orientation will generally increase degree of fill and, therefore, will also increase average retention time. For a given beater configuration, changing the total throughput will generally move the average retention time approximately along one of the fill lines. This is demonstrated by the data points shown in Figure 9 that represent measured average retention time for a given beater configuration when operated at different throughputs.

Retention time distribution is usually expressed in one of two ways. For any sample collected at the preconditioner discharge, the concen-

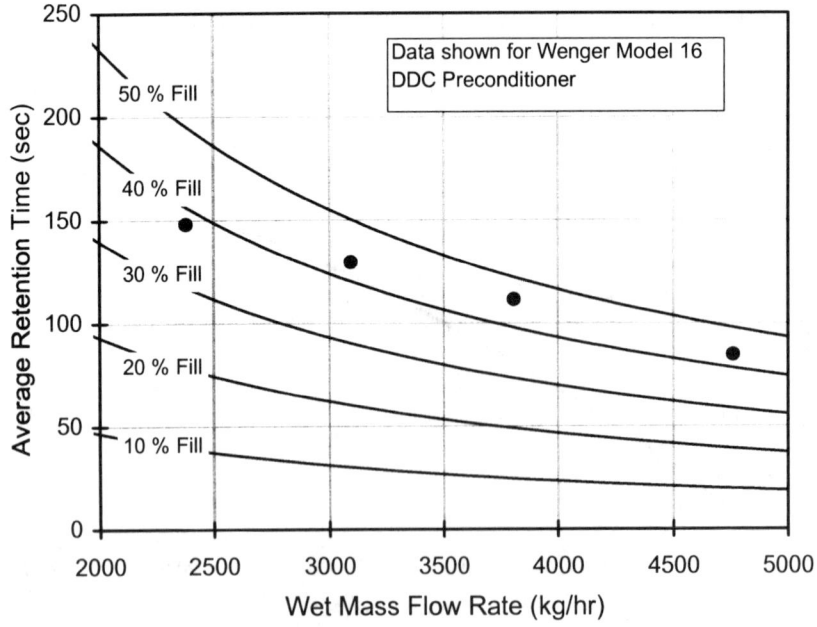

Figure 9 Average retention time map for preconditioners.

tration curve, like the one shown in Figure 10, shows the percent of that sample that was contained for any given time. The accumulation curve, like the one shown in Figure 11, shows the total percent of the sample contained for at least the amount of time read along the x-axis. The average retention time can be found from this chart by reading retention time at the 50% retention level.

Ideally, one would like all raw material particles to be treated in the same fashion to have completely uniform treatment. In addition, the retention time should be long enough to allow for complete penetration of moisture into the particles. This means that the concentration curve should be as narrow as possible and shifted as far to the right as necessary. Similarly, the accumulation curve should be very steep and shifted as far to the right as needed. In practice, the operation variables chosen by the operator affect the shape of these retention time distribution curves.

As already mentioned, increasing the throughput through the preconditioner reduces the average retention time. In addition, increasing throughput will tend to narrow the retention time distribution while shifting it to shorter retention time (Morales et al., 1996). Also, if the preconditioner beaters are not set very close to the inner wall of the pre-

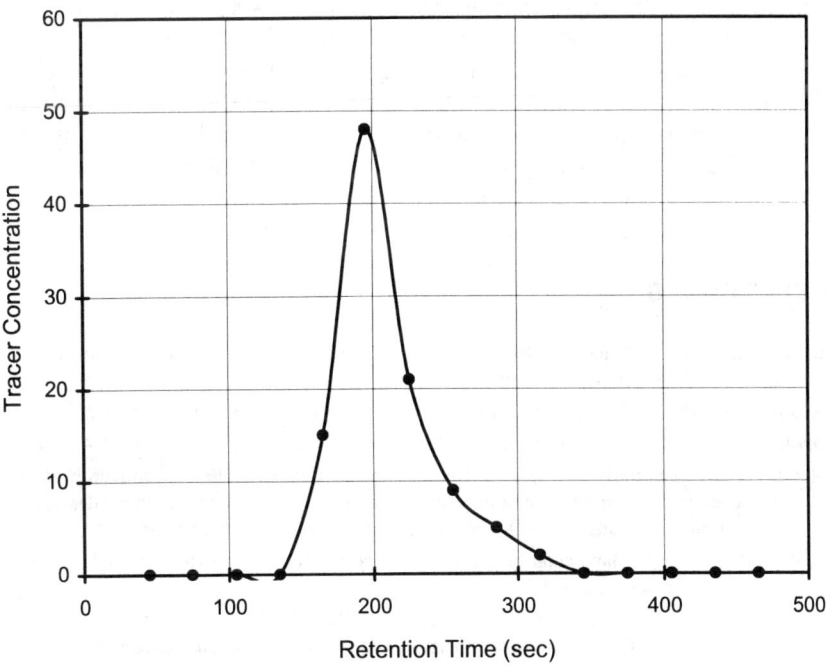

Figure 10 Retention time distribution concentration curve.

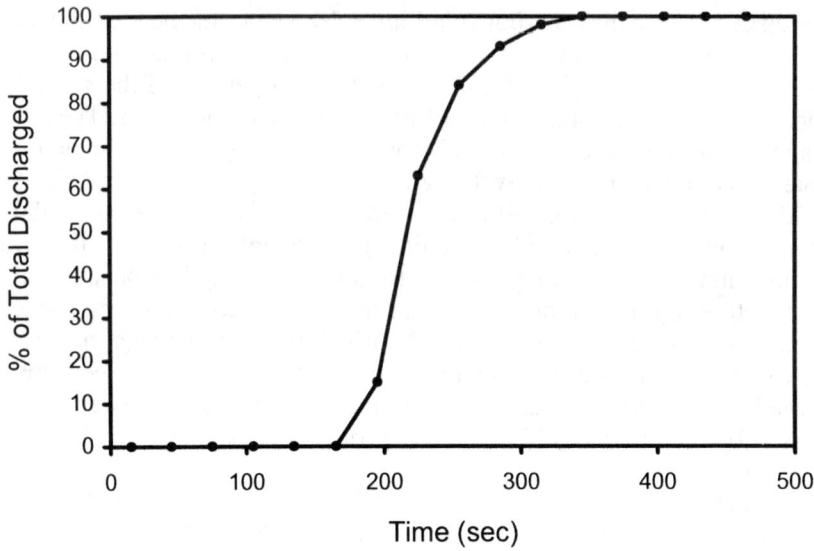

Figure 11 Retention time distribution accumulation curve.

conditioner, the retention time distribution will become very wide and shifted to slightly longer average retention time. A similar effect is seen when the preconditioner shaft speed is reduced. While it may seem that a benefit is obtained from the longer average retention time, this benefit is offset by the reduction in mixing and less uniform retention time.

In conclusion, preconditioning is a very useful tool for increasing the effectiveness of the extrusion process. Through preconditioning, the raw materials being introduced to the extruder can be transformed into a physical form that makes the extruder more efficient and more effective.

REFERENCES

de Muelenaere, H. J. H. and J. L. Buzzard. 1969. "Cooker extruders in service of world feeding." *Food Technology.* 23: 345–351.

Hauck, B. W. 1988. "Preconditioning apparatus for extruder." U.S. Patent No. 4,752,139.

Levine, L. 1995. "Of paddle mixers and preconditioners." *Cereal Foods World.* 40: 452–453.

Morales, J., W. Twombly, J. Brent, B. Strahm, and B. Glaser. 1996. "Examination of the residence time distribution in a differential diameter preconditioner." Presented at the 1996 Annual Meeting of the American Association of Cereal Chemists.

Strahm, B. 1998. "Fundamentals of polymer science as an applied extrusion tool." *Cereal Foods World.* 43: 621–625.

Zhang, W., R. C. Hoseney, and J. M. Faubion. 1998. "Capillary rheometry of corn endosperm: glass transition, flow properties, and melting of starch." *Cereal Chemistry.* 75: 863–867.

CHAPTER 7

Chemical and Nutritional Changes in Food during Extrusion[1]

MARY ELLEN CAMIRE

WITHIN the extruder barrel, unique chemical transformations occur. These changes, coupled with the variable composition of foods, present a significant challenge to food scientists. Although food extrusion borrows from extrusion of synthetic polymers, plastics are much more homogenous and do not present the difficulties encountered with food. Thus, modeling of food extrusion must factor natural variations in moisture, starch, and protein content, as well as any experimental changes. Fortunately, food scientists have a reference collection of more than 500 articles to assist with product and process development.

A few books are focused on extrusion cooking (Frame, 1994; Harper, 1981; Hayakawa, 1992; Kokini et al., 1992; Mercier et al., 1989; O'Connor, 1987). Several reviews of chemical or nutritional changes during extrusion cooking of foods have been published (Björck and Asp, 1983; Camire, 1998; Camire et al., 1990; Cheftel, 1986; de la Gueriviere et al., 1985). Although no journal is dedicated specifically to food extrusion, many relevant papers are published in *Cereal Chemistry*, *Journal of Agricultural and Food Chemistry*, *Journal of Food Engineering*, *Journal of Food Science*, and other food journals. Many

[1]Maine Agricultural and Forest Experiment Station publication #2348.

TABLE 1. General Changes in Foods During Extrusion Cooking.

Chemical Changes	Physicochemical Changes
cleavage	binding of smaller molecules
recombination of fragments	loss of native structure
thermal degradation	losses at die

papers published in the 1970s and 1980s investigated the feasibility of extruding unusual blends of materials, while papers published in the 1990s generally report chemical changes with subsequent physical, sensory, and nutritional effects.

Five general chemical or physicochemical changes can occur during extrusion cooking: binding, cleavage, loss of native conformation, recombination of fragments, and thermal degradation (Table 1). Composition of feed materials is altered by physical losses including leakage of oil and evaporation of water and volatile compounds at the die. Since most chemical reactions occur in the portion of the barrel just before the die, thermally labile compounds such as flavors and vitamins may be injected at that site to minimize exposure to heat and shear.

CRITICAL FACTORS

Typically, extrusion experiments examine just two or three extrusion processing variables, but many factors are important (Table 2). Extruder operators may select parameters for the primary factors, and these factors in turn determine a secondary set of factors: specific mechanical

TABLE 2. Factors that Influence Chemical Changes During Extrusion.

Primary	Secondary
barrel temperature	mass (product) temperature
die geometry	pressure
extruder model	specific mechanical energy
feed composition	
feed moisture	
feed particle size	
feed rate	
screw configuration	
screw speed	

energy (SME), product or mass temperature (PT), and pressure (Meuser and van Lengerich, 1984). These factors influence the viscosity of the food within the extruder barrel, the residence time of the material in the extruder, and the shear applied to the food. Variations caused by feed composition and prior processing of the feed materials are important sources of experimental variation.

The type of extruder used certainly affects chemical reactions. Larger extruders have longer barrels and relatively longer residence times than small tabletop or laboratory-scale extruders. Small extruders require considerably less feed due to the lower throughput, offering a definite advantage for preliminary studies using ingredients available only in small quantities. Many universities that conduct extrusion research have small, single-screw machines. The U.S. food industry, on the other hand, primarily uses large twin-screw extruders, but smaller machines may be used in pilot plants. Improved predictive models would facilitate direct comparison between small research extruders and production-sized equipment, saving both time and money in extrusion research and development.

STARCH

Starchy grains and tubers are the major sources of energy in the diet, particularly for persons living in less-developed nations; therefore, starch changes during extrusion have important nutritional effects. A major difference between extrusion and other forms of food processing is that gelatinization occurs at much lower moisture levels (12–22%). Waxy or high-amylopectin corn starch extrusion is dominated by melting, which follows zero order kinetics, instead of gelatinization, which is a first order reaction (Qu and Wang, 1994).

Gelatinization can be ascertained by a number of methods including thermal data and susceptibility to amylase. Near infrared reflectance spectroscopy (Guy and Osborne, 1996) and the Rapid Viscoanalyzer (Whalen et al., 1997) have been suggested as methods for measuring the extent of cooking in extruded starchy foods. Although complete gelatinization did not occur under the extrusion conditions used, extrusion greatly increased the enzyme digestibility of wheat bran and whole flour starch (Wang et al., 1993). Lipids, sucrose, salt, and dietary fiber modulate gelatinization of starchy foods and may affect expansion and other physical properties (Jin et al., 1994).

The highly branched structure of amylopectin is prone to shear, but amylose and amylopectin molecules may decrease in molecular weight. Larger amylopectin molecules in corn flour had the greatest molecular weight reductions (Politz et al., 1994a). Wheat flour starch molecular weight was best retained by higher die temperature (185°C) and feed moisture (20%) (Politz et al., 1994b). The extent of starch degradation was greater for wheat than for corn. High molecular weight ($>10^7$) starches disappeared during extrusion, and there was a general increase in starch molecules with weights of 10^5–10^7.

Extrusion can be used to direct molecular degradation in order to manufacture dextrins or glucose. High shear conditions are necessary to maximize conversion of starch to glucose. Glucose has been produced by starch extrusion in barley (Linko et al., 1983), cassava (Grossman et al., 1988), corn (van Zuilichem et al., 1990; Roussel et al., 1991), and potato waste (Camire and Camire, 1994). Although enzymes are usually inactivated during extrusion, thermally stable enzymes added to starch prior to extrusion increase the reaction rate within the barrel. Analysis of a response surface experiment revealed that sago starch was most saccharified when a high level of Termamyl, a thermostabile amylase, was added and low moisture conditions were employed (Govindasamy et al., 1995). Conversely, glucose has been mixed with citric acid in a twin-screw extruder to form polydextrose and oligosaccharides with random glucose links (Hwang et al., 1998).

An important consequence of starch degradation is reduced expansion. Expansion, which is usually measured by comparing the diameter of the extrudate to that of the die hole, is related to product texture. Highly expanded products may crumble easily due to thin cell walls, while dense products are often hard. Table 3 lists factors that affect expansion of starchy materials (Chinnaswamy, 1993).

Theoretically, terminal reducing sugars on amylopectin branch fragments could form anhydro links with other carbohydrates. Such novel compounds are likely to be resistant to digestive enzymes. These products are difficult to analyze, but transglycosidation has been reported (Theander and Westerlund, 1987). However, Politz et al. (1994b) did not detect changes in 2,3-glucose linkages in extruded wheat flour after methylation analysis. One explanation for this disagreement between studies is that there are differences in extrusion conditions, including type of extruder used.

TABLE 3. Factors Increasing Expansion of Starchy Materials.

Extrusion Factors	Feed Factors
Barrel temperature (↑)	Moisture content (↓)
Screw configuration	Amylose content (↑)
Die length/diameter ratio (varies)	Dietary fiber, protein, lipid content (generally ↓)

Starch digestibility is largely dependent upon full gelatinization. High starch digestibility is essential for specialized nutritional foods such as infant and weaning foods. Creation of resistant starch by extrusion may have values in reduced calorie products. A patent for production of resistant starch digests high amylose starch with pullulanase, then extrudes the mixture to remove moisture; the process increased resistant starch to 30% (Chiu et al., 1994).

Adding citric acid or high-amylose cornstarch to cornmeal prior to extrusion increased resistant starch plus dietary fiber (Unlu and Faller, 1998), but cost and flavor could be limiting factors. Added dietary fiber also affects digestibility. Cellulose added to cornstarch decreased starch solubility (Chinnaswamy and Hanna, 1991). Longer cellulose fibers produced a greater reduction in solubility, possibly due to transglycosidation.

Complexes of lipids within amylose can be formed during many food processes, including extrusion. The type of starch and lipid present in a food influence the extent to which complexes are formed, with monoglycerides and free fatty acids more likely to form complexes than triglycerides when added to high-amylose starch (Bhatnagar and Hanna, 1994a). In a related study, low feed moisture (19%) and low barrel temperature (110–140°C) resulted in the largest amount of complex formation between stearic acid and normal cornstarch with 25% amylose (Bhatnagar and Hanna, 1994b). High viscosity and longer residence time may favor complex formation. However, this research has not been confirmed with larger twin-screw extruders.

Reactive extrusion is used widely in the plastics industry to produce desirable chemical modifications. A similar approach can be applied to manufacture modified starch. Enzymatic hydrolysis of amioca, a, waxy (high-amylopectin) corn starch results in materials that gel readily. Extruded high amylopectin cornstarch retained most $\alpha(1 \rightarrow 4)$ glycosidic bonds and was highly water-dispersible (Orford et al., 1993). Lupersol 101, a GRAS di-tertiary alkyl peroxide, successfully catalyzed formation of branched polymers of polylactide, a polymer used in biodegradable films (Carlson et al., 1998). This technique has applications for production of novel and starches and gums.

DIETARY FIBER

Numerous compounds are found in plant cell walls, and thus, it is difficult to define, let alone analyze, dietary fiber. Extrusion studies employing different fiber analyses often reach contradictory conclusions. The AOAC total dietary fiber method required for U.S. nutritional labeling measures compounds that are not digested by amylase and protease and are also insoluble in 80% aqueous ethanol. Compounds modified during extrusion may become indigestible and, thus, have similar properties to fiber in the body.

Artz and coworkers (1990) measured X-ray diffraction patterns of corn fiber-cornstarch blends before and after extrusion and found no change in patterns due to extrusion. Since the corn bran isolate contained only 16.6% crystalline cellulose, X-ray diffraction may not indicate change in other fiber constituents such as hemicelluloses.

Measurement of total dietary fiber may meet food labeling requirements, but this assay also does not detect changes in fiber solubility induced by extrusion. In our laboratory, an enzymatic-chemical method discerned extrusion changes in lignin and nonstarch polysaccharides (NSP), but extrusion did not affect uronic acids in the foods studied (Camire and Flint, 1991). Soluble NSP was higher after extrusion in oatmeal and potato peels; no difference was found in cornmeal.

Extrusion reduced pectin and hemicellulose molecular weight, resulting in a threefold increase in water solubility of sugar beet pulp fiber (Ralet et al., 1991). Although grinding increased pea hull soluble fiber to 8% (dry basis), extrusion resulted in soluble fiber levels over 10% (Ralet et al., 1993). However, total fiber was lower after extrusion. Smaller molecules are more soluble in water and may also be soluble in aqueous ethanol, which is used for extraction in enzymatic-gravimetric and enzymatic-chemical methods of dietary fiber analysis. Acid and alkali somewhat raised soluble fiber in corn bran (Ning et al., 1991).

Extrusion significantly increased soluble fiber in beans (cv. Horsehead) (Figure 1), but insoluble fiber was decreased only in samples extruded at the lower moisture content used in the study (Martín-Cabrejas et al., 1999). The additional soluble fiber was composed of xylose, arabinose, mannose, and uronic acids, but glucose from the insoluble fiber in samples extruded at 25% moisture was not accounted for. No clear trend for Klason lignin changes was apparent.

The advantage of increased soluble fiber in foods is due to health benefits attributed to soluble fiber. It is not clear whether the soluble fiber created during extrusion has the same health effects as naturally soluble materials. Research studies have revealed contradictory findings. In an *in vitro* assay that mimics the digestive system, a significantly greater amount of cholic acid and deoxycholic acid were bound by potato peels extruded at a lower barrel temperature (Camire et al., 1994). These peels also had higher soluble fiber.

Extruded grains contained higher soluble fiber, and soluble β-glucans increased slightly in extruded barley and oats (Wang and Klopfenstein, 1993). Lower total serum and liver cholesterol levels were found

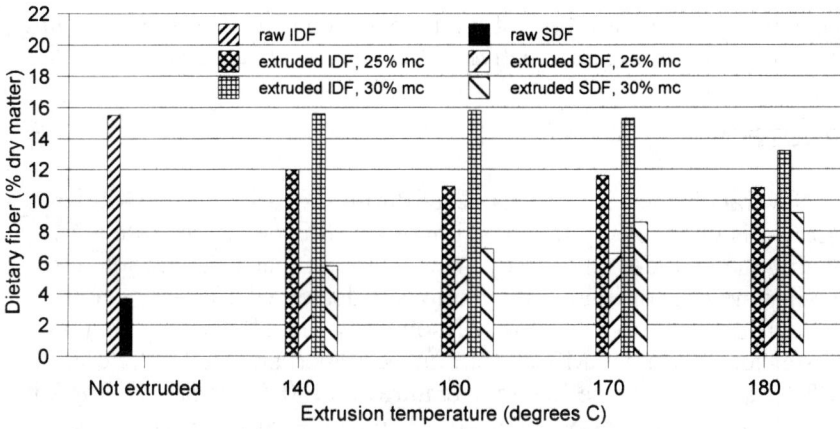

Figure 1 Changes in insoluble dietary fiber (IDF) and soluble dietary fiber (SDF) in beans (*Phaseolus vulgaris* L., cv. Horsehead) before and after extrusion. Extrusion was performed at four different barrel temperatures (140, 160, 170, and 180°C) and two feed moisture contents (25% and 30%). [Adapted from Table 2, Martín-Cabrejas et al., 1999. "Modifications to physicochemical and nutritional properties of hard-to-cook beans (*Phaseolus vulgaris* L.) by extrusion cooking." *J. Agric. Food Chem.* 47: 1174–1182.]

in young rats fed extruded barley, oats, and wheat, compared with levels in rats fed a standard diet or unextruded grains. The viscosity of aqueous suspensions of extruded grains were highest. Low-energy extrusion significantly altered wheat bran as evidenced by lower serum and liver cholesterol and total liver lipids in rats fed the brans (Kahlon et al., 1998). The process had no effect on corn, oat, or rice brans. Viscous gums and other soluble fibers may reduce cholesterol levels by trapping bile acids; increased excretion of bile eventually depletes body stores of cholesterol, which are tapped to synthesize new bile acids.

Viscous gels formed by soluble fiber in the small intestine slow the absorption of glucose, preventing peaks in serum glucose levels. Higher levels of soluble fiber in lemon and orange peels after extrusion were associated with increased *in vitro* viscosity (Gourgue et al., 1994), starch digestion and diffusion of glucose were not affected by extrusion.

Insoluble dietary fiber also has health benefits. One possible effect is protection against colorectal cancer by binding dietary carcinogens. Although other extrusion conditions did not affect the ability of potato peels to bind the polycyclic aromatic hydrocarbon benzópyrene, extrusion conditions of 110°C barrel temperature and 30% feed moisture significantly reduced the ability of potato peels to bind the carcinogen (Camire et al., 1995b). In a related study, sixteen extruded commercial cereals bound at least 40% of the benzópyrene added at the beginning of *in vitro* digestion, irrespective of fiber content (Camire et al., 1995c).

PROTEIN

Several changes in proteins occur during extrusion (Table 4); denaturation is undoubtedly the most important. Most enzymes lose activity within the extruder unless they are stable to heat and shear. Protein solubility in water or dilute salt solutions is decreased after extrusion. Although denaturation and loss of solubility are affected by increased barrel temperature, SME appears to be important (Della Valle et al., 1994). Even under the low temperatures ($<100°C$) of pasta extrusion, wheat protein solubility is reduced (Ummadi et al., 1995a).

The subject of extrusion effects on protein has been addressed in reviews by Arêas (1992) and Camire (1991). During extrusion, disulfide bonds are broken and may re-form. Electrostatic and hydrophobic interactions favor formation of insoluble aggregates. The creation of new peptide bonds during extrusion is controversial. High molecular weight proteins can dissociate into smaller subunits. Exposure of enzyme-susceptible sites improves digestibility.

Although many researchers have used extrusion temperatures below 150°C, different mechanisms might occur at higher temperatures. Prudîncio-Ferreira and Arêas (1993) extruded soy protein at three barrel temperatures (140, 160, and 180°C) and two feed moistures (30 and 40%). Insolubility due to disulfide bonds was highest in soy extruded at 40% feed moisture but decreased with increasing barrel temperature at both moisture levels.

Maillard reactions occur during extrusion, particularly at higher barrel temperatures and lower feed moistures. Free sugars may be produced during extrusion to react with lysine and other amino acids with free terminal amines. Starch and dietary fiber fragments as well as sucrose hydrolysis products are available for Maillard reactions. Lower pH favored Maillard reactions, as measured by increased color in a model

TABLE 4. **Protein Changes During Extrusion.**

Functional Changes	Nutritional Changes
Reduced solubility in water and diluted buffer	Reduced lysine
Texturization	Improved digestibility

system consisting of wheat starch, glucose, and lysine (Bates et al., 1994).

The nutritional significance of Maillard reactions is most important for animal feeds and foods intended for special nutritional needs such as weaning or intended as the sole item in a diet. Since many extruded foods in the U.S. are not major sources of protein, the losses of lysine and other essential amino acids have little impact on nutrition. In a rat feeding study, extrusion texturized soy isolate did not change rat serum cholesterol, fecal steroid excretion, protein digestibility, or biological value compared with nonextruded soy (Fukui et al., 1993).

One of the first major applications for extruders was production of meat analogues in the 1960s. Fibrous texture can be developed during extrusion that mimics the texture of meat. Soy protein should be denatured within the extruder at a pH near the soy isoelectric point to achieve adequate texture formation. Added sodium hydroxide (or any alkali) is detrimental to texturization (Dahl and Villota, 1991). A novel application for soy texturization was developed by Huang and colleagues (1995) at Iowa State University. They developed a process to extrude soy protein isolate into textile fibers. One problem was brittleness of the fibers. The addition of glycerol during extrusion or chemical treatments postextrusion reduced brittleness.

High-moisture extrusion has been used to create novel foods and ingredients such as gels and emulsions (Cheftel et al., 1992). This field is relatively new, and few papers have been published on the subject. Barraquio and van de Voort (1991) employed extrusion to acid coagulate skim milk powder, then neutralized the acid casein to sodium caseinate during a second extrusion step. Another dairy by-product, whey protein isolate, was extruded at low pH, barrel temperature, and screw speed, resulting in coagulated semisolid spreads that the researchers proposed could act as fat substitutes (Queguiner et al., 1992).

LIPIDS

High-fat materials are generally not extruded. Lipid levels over 5–6% impair extruder performance. Torque is decreased because the lipid reduces slip within the barrel, and often product expansion is poor because insufficient pressure is developed during extrusion. Lipid is released from cells due to the cooking and physical disruption of plant cell walls.

In general, food lipid content appears to be lower after extrusion. Some lipid may be lost at the die as free oil, but this situation only occurs with high-fat materials such as whole soy. Another explanation for the lower lipid level is formation with complexes with amylose or protein. When extrudates are digested with acid or amylase, then extracted with solvent, lipid recovery is higher. Although only half of the ether-extractable lipids in extruded whole wheat was recovered, acid hydrolysis indicated that total fat was not significantly changed due to extrusion (Wang et al., 1993). Wheat bran, which has less starch than the whole grain, had more free lipids after extrusion, presumably because lipid-amylose complexes could not form.

Free fatty acids are a problem in foods because they are more susceptible to oxidation than are triglycerides. Free fatty acids also produce off-flavors. Although the hydrolysis of triglycerides to glycerol and free fatty acids is theoretically possible, this reaction does not appear to occur to a significant extent. In fact, extrusion can prevent free fatty acid release by denaturing hydrolytic enzymes. This activity is exploited in commercial processes to stabilize rice bran.

Lipid oxidation can rapidly deteriorate sensory and nutritional qualities of foods and feeds. Lipid oxidation probably does not take place during extrusion due to the short residence time for most extrusion processes. However, rancidity is a concern for extruded products during storage. Table 5 summarizes factors that influence lipid oxidation in extruded products. Screw wear is a concern since metals can act as pro-oxidants. Iron content and peroxide values were higher in extruded rice and dhal compared with similar products processed by drying methods (Semwal et al., 1994).

The larger surface area created by the air cells throughout highly expanded extrudates favors oxidation. On the other hand, extrusion denatures enzymes that can promote oxidation, and lipids held within starch are less susceptible to oxidation. Compounds produced by Maillard reactions can also act as antioxidants. Oatmeal cookies with added potato peels had lower peroxide values than control samples, and higher antioxidant activity was observed for extruded peels compared to unextruded peels (Arora and Camire, 1994).

TABLE 5. Extrusion Factors that Affect Lipid Oxidation.

Factors that Increase Oxidation	Factors that Reduce Oxidation
Screw wear	Enzyme inactivation
Expansion	Formation of antioxidant compounds via Maillard reactions
Low water activity	Lipid-amylose complexes

VITAMINS

Since vitamins differ greatly in composition, their stability during extrusion is also variable. This subject is addressed in reviews on nutritional changes during extrusion (Björck and Asp, 1983; Camire, 1998; Camire et al., 1990; Cheftel, 1986; de la Gueriviere, et al., 1985) and in a review of vitamin retention by Killeit (1994) (Table 6). Minimizing temperature and shear within the extruder protects most vitamins.

Among the lipid-soluble vitamins, vitamins D and K are fairly stable. Vitamins A and E and their related compounds, carotenoids and tocopherols, respectively, are not stable in the presence of oxygen and heat. The vitamin A precursor β-carotene is added to foods as a coloring agent and antioxidant. Thermal degradation appears to be the major factor contributing to β-carotene losses during extrusion. Higher barrel temperatures (200°C compared with 125°C) reduced all trans β-carotene in wheat flour by over 50% (Guzman-Tello and Cheftel, 1990). Oxidation contributed to the loss, since added BHT or extrusion under nitrogen resulted in smaller pigment losses.

Grains, including those used in extruded foods, must be enriched with B vitamins in the United States. Very little is known about the relative extrusion stability of added nutrients compared with naturally occurring vitamins. The water soluble vitamin most susceptible to thermal processing is thiamine. Thiamine stability during extrusion is highly variable, as evidenced by Killeit's (1994) report that losses range between 5–100%. For example, Andersson and Hedlund (1990) observed high losses of thiamine when no water was added during extrusion, but riboflavin (B_2) and niacin were not affected.

Ascorbic acid (vitamin C) is also sensitive to heat and oxidation. This vitamin decreased in wheat flour when extruded at higher barrel temperatures at fairly low moisture (10%) (Andersson and Hedlund, 1990).

TABLE 6. Extrusion Parameters that Increase Vitamin Destruction.[1]

↑ barrel and mass temperatures
↑ screw speed
↑ specific energy input
↓ feed moisture
↓ die diameter
↓ throughput

[1]Adapted from Killeit, 1994.

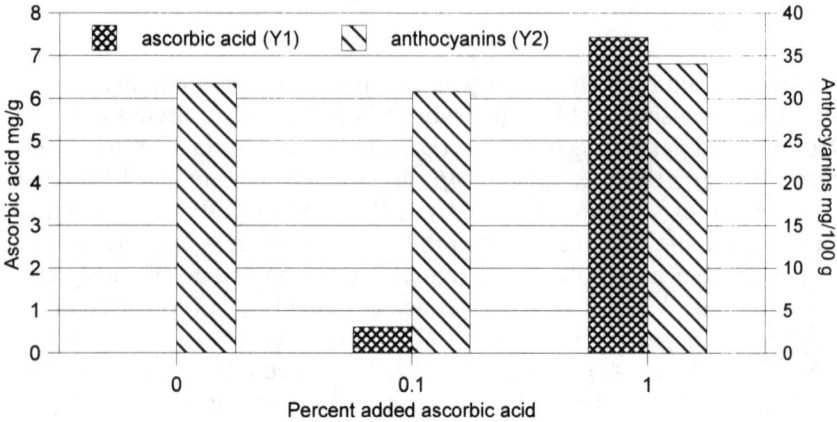

Figure 2 Retention of ascorbic acid and anthocyanins in a model corn cereal after extrusion in the presence of added ascorbic acid.

Added vitamin C was fairly stable in a model extruded breakfast cereal containing blueberry concentrate (Figure 2), but losses were greater in similar products without blueberry concentrate (Chaovanalikit and Camire, unpublished data).

MINERALS

Although essential for health, minerals stability during extrusion has been studied little because they are stable in other food processes. Research has focused on two general topics: binding of minerals by fiber and other macromolecules and adding of minerals due to screw and barrel wear.

During the late 1970s, nutritionists became concerned that high-fiber diets might impair mineral absorption. Although these concerns now seem to be unfounded for most Americans, phytate in whole grains chelates some minerals. Extrusion can affect mineral absorption by altering phytate. Extrusion reduced phytate levels in wheat flour (Fairweather-Tait et al., 1989) but not in legumes (Lombardi-Boccia et al., 1991). Boiled legumes and ones extruded under high-shear conditions had less dialyzable iron than did samples extruded under low-shear conditions (Ummadi et al.,1995b). Although phytic acid was lower under all processing conditions, total phytate was not affected.

Foods with higher dietary fiber content increase the transfer of metal from the extruder screws and barrel to the food itself. Iron levels in potato peels increased by at least 38% as a result of extrusion, and the greatest levels were found in samples extruded at the higher barrel temperature (Camire et al., 1994). Screw wear iron was well-utilized by rats fed extruded corn and potato (Fairweather-Tait et al., 1987). Adults fed extruded wheat bran and wheat flour absorbed iron and zinc as well as they did from the nonextruded materials (Fairweather-Tait et al., 1989).

Fortification of foods with minerals prior to extrusion poses other problems. Iron forms complexes with phenolic compounds that are dark in color and detract from the appearance of foods. Ferrous sulfate heptahydrate was found to be a suitable source of iron in a simulated rice product because it did not discolor (Kapanidis and Lee, 1996). Added calcium hydroxide (0.15–0.35%) decreased expansion and increased lightness in color of cornmeal extrudates (Martínez-Bustos et al., 1998).

PHYTOCHEMICALS

Extrusion research is just now providing clues as to the fate of nutrients during extrusion. As nutrition science begins to unravel the importance of nonnutrient chemicals in foods, it is clear that extrusion effects on these compounds must be studied. For example, genistein and phytoestrogens in soy may help prevent breast cancer, yet extrusion texturization of soy to produce more palatable soy foods might significantly reduce these compounds. Extrusion of soy protein concentrate and a mixture of cornmeal and soy protein concentrate (80:20) did not result in changes in total isoflavone content (Mahungu et al., 1999). The aglycones and malonyl forms tended to decrease with extrusion, while acetyl derivatives increased. The nutritional implications of such changes are not known.

Phenolic compounds in grains, fruits, and vegetables act as antioxidants and may have health benefits. Total free phenolics, primarily chlorogenic acid, decreased significantly due to extrusion in potato peels produced by steam peeling (Figure 3, Camire and Dougherty, unpublished data). More phenolics were retained with higher barrel temperature and feed moisture. We suspect that the lost phenolics reacted with themselves or with other compounds to form larger insoluble materials. In a model breakfast cereal containing cornmeal and sucrose, anthocyanin pigments were degraded at higher levels of added ascorbic acid (Figure 2), and total anthocyanins were significantly decreased by extrusion (Chaovanalikit and Camire, unpublished data). Polymerization appears to have contributed to anthocyanin losses as well. Within the next five years, hopefully this phenomenon and others will be more fully understood.

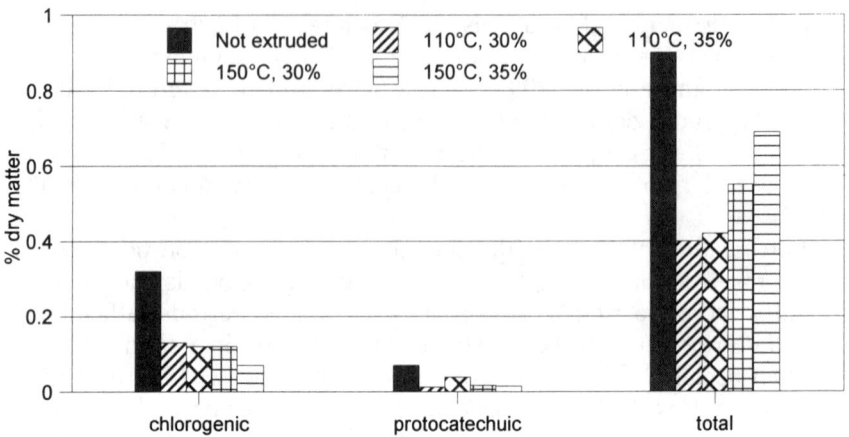

Figure 3 Individual and total phenolics in extruded potato peels processed at two different barrel temperatures (110 or 150°C) and feed moisture (30 or 35%).

NATURAL TOXINS

One of the most important benefits of extrusion cooking is the reduction of natural toxins and antinutrients since many foods, especially legumes, have chemicals that are poisonous or reduce utilization of nutrients. Although some of these undesirable compounds may have health benefits for adults, children and young animals can suffer retarded growth. Recent research on this subject has focused on the application of chemical treatments in combination with extrusion cooking.

Many foods, particularly legumes, contain trypsin inhibitors (TI) that interfere with enzymes that digest protein. Long-term consumption of TI leads to impaired growth and to pancreatic hypertrophy, since the body responds to TI by producing more enzymes. In order to attain the same nutritional quality for chicks as soybeans without the Kunitz TI, common soybeans had to be extruded at higher temperatures (138–154°C versus 121–138°C) (Zhang et al., 1993). Van den Hout and coworkers (1998) concluded that heat, not shear, was primarily responsible for TI inactivation in soy flour.

Another important safety issue is food allergens. Although extrusion cannot eliminate allergenic proteins in foods, promising developments may lead to significant reductions. Using kneading disc screw elements permits allergen reductions in soy at relatively low temperatures, presumably by inducing denaturation with the increased shear (Ohishi et al., 1994). Increasing barrel temperature also reduced allergenicity, but alterations in feed rate and screw speed were unsuccessful. Product temperature and SME were important factors in the loss of protein antigenicity, as measured by immunoassay, in extruded pea flour (Della Valle et al., 1994).

Fortunately, several natural toxins may be reduced simultaneously with extrusion. Lectin and α-amylase inhibitor were completely eradicated in extruded bean samples, with the exception of beans processed at 140°C and 30% feed moisture (Martín-Cabrejas et al., 1999). TI was significantly reduced in all extruded beans as well. Partial reductions in TI activity and almost complete destruction of hemagglutinins were found in extruded starch fractions of beans (Gujska and Khan, 1991). Extruded jack beans had significantly lower levels of hemagglutinins and urease activity; canavanine was not affected by extrusion (Tepal et al., 1994). Screw speed changes were not effective in this study. Nitrogen solubility was also lower in the extruded beans and may have the potential to serve as a marker for toxin inactivation since Kjeldahl analyses are less expensive than many of the tests for toxins.

Neither α-chaconine or α-solanine, the major glycoalkaloids in potatoes, were reduced during twin-screw extrusion of potato peels, but just 3–5% of the glycoalkaloids were soluble during in vitro digestion (Zhao and Camire, 1994). Solubility is essential for absorption, so extrusion may reduce the toxicity of these compounds without destroying them. Using the same extrusion conditions, potato TI were significantly reduced in potato peels produced by abrasion (Zhao and Camire, 1995). Very little has been published on the effects of extrusion on pesticides, and potato peels are an excellent model since tubers are treated with several chemicals during storage. Unfortunately, neither chlorpropham, a sprouting inhibitor, nor thiabendazole, a fungicide, were significantly reduced by extrusion (Camire et al., 1995a).

Other natural toxins have been studied for their susceptibility to extrusion. Added ammonia in combination with extrusion decreased total glucosinolates in canola and weeds, but such treatment would not be suitable for foods destined for human consumption due to the residual ammonia (Darroch et al., 1990). Alkylresorcinols decreased after extrusion by nearly 50%; variations in extrusion parameters had no impact on this reduction (Al-Ruqaie and Lorenz, 1992). Although not considered toxic, raffinose and stachyose are undesirable in foods because they promote flatulence. Avoidance of healthy foods containing these oligosaccharides could impair the nutritional status of some consumers. Higher barrel temperature was most effective in reducing raffinose and stachyose in high-starch fractions of pinto beans (Borejszo and Khan, 1992).

FLAVORS

Many extruded foods are bland since there is little time for flavor development. Thermal degradation can occur. Volatile flavors flash off with water vapor when the food exits the extruder at the die. Two recent reviews of extrusion and flavor have been published (Camire and Belbez 1996; Riha and Ho, 1998).

Nonvolatile Maillard reaction compounds have been identified from an extruded model system consisting of wheat starch, lysine, and glucose (Ames et al., 1997). The major volatiles recovered from corn flour extruded at low temperatures and high moisture were derived from lipid reactions (Bredie et al., 1998). Higher barrel temperatures and lower feed moistures favored formation of Maillard compounds, and heterocylic chemicals formed during the process are believed to be important for typical cooked grain flavor.

Postextrusion flavoring is often used to improve acceptability, and some research has evaluated the effects of adding flavoring before or during extrusion on maximizing flavor retention. Retention of added flavor compounds was highest in starch extrudates when the flavors were directly injected into the extruder barrel just before the die (Kollengode and Hanna, 1997a). Low feed moisture (9% versus 17%) and the use of raw starch instead of pregelatinized starch also improved flavor retention. Kollengode and Hanna (1997b) also obtained recoveries of 70–100% when flavors were complexed with cyclodextrins prior to extrusion. The relatively high cost of cyclodextrins may be offset by the reduced cost of adding flavors.

FUTURE DIRECTIONS

Improved predictive models are needed but any model should include food composition and prior processing history. Food scientists and engineers should focus on the relationships between composition changes and product quality, both nutritional and sensory. Very different mechanisms may occur during high-moisture extrusion, creating a new line of research objectives.

REFERENCES

Al-Ruqaie, I. and K. Lorenz. 1992. "Alkylresorcinols in extruded cereal brans." *Cereal Chem.* 69: 472–475.

Ames, J. A., A. Arnoldi, L. Bates, and M. Negroni. 1997. "Analysis of the methanol-extractable nonvolatile Maillard reaction products of a model extrusion-cooked cereal product." *J. Agric. Food Chem.* 45: 1256–1263.

References

Andersson, Y. and B. Hedlund. 1990. "Extruded wheat flour: correlation between processing and product quality parameters." *Food Qual. Prefer.* 2: 201–216.

Arêas, J. A. G. 1992. "Extrusion of food proteins." *Crit. Rev. Food Sci. Nutr.* 32: 365–392.

Arora, A. and M. E. Camire. 1994. "Performance of potato peels in muffins and cookies." *Food Res. Intl.* 27: 14–22.

Artz, W. E., C. C. Warren, and R. Villota. 1990. "Twin screw extrusion modification of corn fiber." *J. Food Sci.* 55: 746–750, 754.

Barraquio, V. L. and F. R. van de Voort. 1991. "Sodium caseinate from skim milk powder by extrusion processing: physicochemical and functional properties." *J. Food Sci.* 56: 1552–1556, 1561.

Bates, L., J. M. Ames, and D. B. MacDougall. 1994. "The use of a reaction cell to model the development and control of colour in extrusion cooked foods." *Lebensm.-Wiss. u. Technol.* 27: 375–379.

Bhatnagar, S. and M. A. Hanna. 1994a. "Amylose-lipid complex formation during single-screw extrusion of various corn starches." *Cereal Chem.* 71: 582–587.

Bhatnagar, S. and M. A. Hanna. 1994b. "Extrusion processing conditions for amylose-lipid complexing." *Cereal Chem.* 71: 587–593.

Björck, I. and N.-G. Asp. 1983. "The effects of extrusion cooking on nutritional value—a literature review." *J. Food Eng.* 2: 281–308.

Borejszo, Z. and K. Khan. 1992. "Reduction of flatulence-causing sugars by high temperature extrusion of pinto bean high starch fractions." *J. Food Sci.* 57: 771–772, 777.

Bredie, W. L. P., D. S. Mottram, and R. C. E. Guy. 1998. "Aroma volatiles generated during extrusion cooking of maize flour." *J. Agric. Food Chem.* 46: 1479–1487.

Camire, M. E. 1991. "Protein functionality modification by extrusion cooking." *J. Am. Oil Chem. Soc.* 68: 200–205.

Camire, M. E. 1998. "Chemical changes during extrusion cooking: recent advances." In: *Process-Induced Chemical Changes in Foods*, F. Shahidi, C.-T. Ho, and N. van Chuyen, eds., New York: Plenum Press, pp. 109–122.

Camire, M. E. and E. O. Belbez. 1996. "Flavor formation during extrusion cooking." *Cereal Foods World.* 41: 734–736.

Camire, M. E., R. J. Bushway, J. Zhao, B. Perkins, and L. R. Paradis. 1995a. "Fate of thiabendazole and chlorpropham residues in extruded potato peels." *J. Agric. Food Chem.* 43: 495–497.

Camire, M. E. and A. L. Camire. 1994. "Enzymatic starch hydrolysis of extruded potato peels." *Starch/Stärke.* 46: 308–311.

Camire, M. E., A. L. Camire, and K. Krumhar. 1990. "Chemical and nutritional changes." *Crit. Rev. Food Sci. Nutr.* 29: 35–57.

Camire, M. E. and M. P. Dougherty, unpublished data.

Camire, M. E. and S. I. Flint. 1991. "Thermal processing effects on dietary fiber composition and hydration capacity in corn meal, oat meal, and potato peels." *Cereal Chem.* 68: 645–647.

Camire, M. E., J. Zhao, M. P. Dougherty, and R. J. Bushway. 1995b. "*In vitro* binding of benzópyrene by extruded potato peels." *J. Agric. Food Chem.* 43: 970–973.

Camire, M. E., J. Zhao, M. P. Dougherty, and R. J. Bushway. 1995c. "*In vitro* binding of benzópyrene by ready-to-eat breakfast cereals." *Cereal Foods World.* 40: 447–450.

Camire, M. E., J. Zhao, and D. A. Violette. 1994. "*In vitro* binding of bile acids by extruded potato peels." *J. Agric. Food Chem.* 41: 2391–2394.

Carlson, D., P. Duboise, L. Nie, and R. Narayan. 1998. "Free radical branching of polylactide by reactive extrusion." *Polymer Engin. Sci.* 38: 311–321.

Chaovanalikit, A. and M. E. Camire, unpublished data.

Cheftel, J. C. 1986. "Nutritional effects of extrusion cooking." *Food Chem.* 20: 263–283.

Cheftel, J. C., M. Kitagawa, and C. Queguiner. 1992. "New protein texturization processes by extrusion cooking at high moisture levels." *Food Rev. Intl.* 8: 235–275.

Chinnaswamy, R. 1993. "Basis of cereal starch expansion." *Carbohydrate Polymers.* 21: 157–167.

Chinnaswamy, R. and M. A. Hanna. 1991. "Physicochemical and macromolecular properties of starch-cellulose fiber extrudates." *Food Structure.* 10: 229–239.

Chiu, C.-W., M. Henley, and P. Altieri. 1994. "Process for making amylase resistant starch from high amylose starch." U.S. Patent 5,281,276, Jan. 25.

Dahl, S. R. and R. Villota. 1991. "Twin-screw extrusion texturization of acid and alkali denatured soy proteins." *J. Food Sci.* 56: 1002–1007.

Darroch, C. S., J. M. Bell, and K. O. Keith. 1990. "The effects of moist heat and ammonia on the chemical composition and feeding value of extruded canola screenings for mice." *Can. J. Anim. Sci.* 70: 267–277.

de la Gueriviere, J. F., C. Mercier, and L. Baudet. 1985. "Incidences of extrusion-cooking on certain nutritional parameters of food products, especially cereals." *Cah. Nutr. Diet.* 20: 201–210.

Della Valle, G., L. Quillien, and J. Gueguen. 1994. "Relationships between processing conditions and starch and protein modifications during extrusion-cooking of pea flour." *J. Sci. Food Agric.* 64: 509–517.

Fairweather-Tait, S. J., D. E. Portwood, L. L. Symss, J. Eagles, and M. J. Minski. 1989. "Iron and zinc absorption in human subjects from a mixed meal of extruded and nonextruded wheat bran and flour." *Am. J. Clin. Nutr.* 49: 151–155.

Fairweather-Tait, S. J., L. S. Symss, A. C. Smith, and I. T. Johnson. 1987. "The effect of extrusion cooking on iron absorption from maize and potato." *J. Sci. Food Agric.* 39: 341–348.

Frame, N. D. 1994. *The Technology of Extrusion Cooking.* Glasgow: Blackie Academic & Professional.

Fukui, K., T. Aoyama, Y. Hashimoto, and T. Yamamoto. 1993. "Effect of extrusion of soy protein isolate on plasma cholesterol level and nutritive value of protein in growing male rats." *J. Jap. Soc. Nutr. Food Sci.* 46: 211–216.

Gourgue, C., M. Champ, F. Guillon, and J. Delort-Laval. 1994. "Effect of extrusion-cooking on the hypoglycaemic properties of citrus fibre: an *in vitro* study." *J. Sci. Food Agric.* 64: 493–499.

Govindasamy, S., O. H. Campanella, and C. G. Oates. 1995. "Influence of extrusion variables on subsequent saccharification behavior of sago starch." *Food Chem.* 54: 289–296.

Grossman, M. V. E., A. A. El-Dash, and J. F. Carvalho. 1988. "Extrusion cooking of cassava starch for ethanol production." *Starch/Stärke.* 40: 300–307.

Gujska, E. and K. Khan. 1991. "Feed moisture effects on functional properties, trypsin inhibitor, and hemagglutinating activities of extruded bean high starch fractions." *J. Food Sci.* 54: 443–447.

Guy, R. C. E. and B. G. Osborne. 1996. "The application of near infrared reflectance spectroscopy to measure the degree of processing in extrusion cooking processes." *J. Food Engin.* 27: 241–258.

Guzman-Tello, R. and J. C. Cheftel. 1990. "Colour loss during extrusion cooking of beta-carotene-wheat flour mixes as an indicator of the intensity of thermal and oxidative processing." *Intl. J. Food Sci. Technol.* 25: 420–434.

Harper, J. M. 1981. *Extrusion of Foods.* Boca Raton, FL: CRC Press, Inc.

Hayakawa, I. 1992. *Food Processing by Ultra High Pressure Twin Screw Extrusion.* Lancaster, PA: Technomic Publ. Co.

Huang, H. C., E. G. Hammond, C. A. Reitmeier, and D. J. Myers. 1995. "Properties of fibers produced from soy protein isolate by extrusion and wet spinning." *J. Am. Oil Chem. Soc.* 72:1453–1460.

Hwang, J.-K., C.-J. Kim, and C.-T. Kim. 1998. "Production of glucooligosaccharides and polydextrose by extrusion reactor." *Starch/Stärke.* 50: 104–107.

Jin, Z., F. Hsieh, and H. E. Huff. 1994. "Extrusion cooking of corn meal with soy fiber, salt, and sugar." *Cereal Chem.* 71: 227–234.

Kahlon, T. S., R. H. Edwards, and F. I. Chow. 1998. "Effect of extrusion on hypocholesterolemic properties of rice, oat, corn, and wheat bran diets in hamsters." *Cereal Chem.* 75: 897–903.

Kapanidis, A. N. and T.-C. Lee. 1996. "Novel method for the production of color-compatible ferrous sulfate-fortified simulated rice through extrusion." *J. Agric. Food Chem.* 44: 522–525.

Killeit, U. 1994. "Vitamin retention in extrusion cooking." *Food Chem.* 49: 149–155.

Kokini, J. L., C.-T. Ho, and M. V. Karwe. (Eds.) 1992. *Food Extrusion Science and Technology.* New York: Marcel Dekker.

Kollengode, A. N. R. and M. A. Hanna. 1997a. "Flavor retention in pregelatinized and internally flavored starch extrudates." *Cereal Chem.* 74: 396–399.

Kollengode, A. N. R. and M. A. Hanna. 1997b. "Cyclodextrin complexed flavors retention in extruded starches." *J. Food Sci.* 62: 1057–1060.

Linko, P., S. Hakulin, and Y.-Y. Linko. 1983. "Extrusion cooking of barley starch for the production of glucose syrup and ethanol." *J. Cereal Sci.* 1: 275–284.

Lombardi-Boccia, G., G. Di Lullo, and E. Carnovale. 1991. "*In vitro* iron dialysability from legumes: influence of phytate and extrusion cooking." *J. Sci. Food Agric.* 55: 599–605.

Mahungu, S. M., S. Diaz-Mercado, J. Li, M. Schwenk, K. Singletary, and J. Faller. 1999. "Stability of isoflavones during extrusion processing of corn/soy mixture." *J. Agric. Food Chem.* 47: 279–284.

Martín-Cabrejas, M. A., L. Jaime, C. Karanja, A. J. Downie, M. L. Parker, F. J. Lopez-Andreu, G. Maina, R. M. Esteban, A. C. Smith, and K. W. Waldron. 1999. "Modifications to physicochemical and nutritional properties of hard-to-cook beans (*Phaseolus vulgaris* L.) by extrusion cooking." *J. Agric. Food Chem.* 47: 1174–1182.

Martínez-Bustos, F., Y. K. Chang, A. C. Bannwart, M. E. Rodríguez, P. A. Guedes, and E. R. Gaiotti. 1998. "Effects of calcium hydroxide and processing conditions on corn meal extrudates." *Cereal Chem.* 75: 796–801.

Mercier, C., P. Linko, and J. M. Harper. (Eds.) 1989. *Extrusion Cooking.* Am. Assoc. Cereal Chem., St. Paul, MN.

Meuser, F. and B. van Lengerich. 1984. "Systems analytical model for the extrusion of starches." In: *Thermal Processing and Quality of Foods*, P. Zeuthen, J. C. Cheftel, C. Eriksson, M. Jul, H. Leniger, P. Linko, G. Varela, and G. Vos, eds. London: Elsevier Applied Sci. Publ. p. 175–179.

Ning, L., R. Villota, and W. E. Artz. 1991. "Modification of corn fiber through chemical treatments in combination with twin-screw extrusion." *Cereal Chem.* 68: 632–636.

O'Connor, C. (Ed.) 1987. *Extrusion Technology for the Food Industry.* London: Elsevier Applied Sci. Publ.

Ohishi, A., K. Watanabe, M. Urushibata, K. Utsuno, K. Ikuta, K. Sugimoto, and H. Harada. 1994. "Detection of soybean antigenicity and reduction by twin-screw extrusion." *J. Am. Oil Chem. Soc.* 71: 1391–1396.

Orford, P. D., R. Parker, and S. G. Ring. 1993. "The functional properties of extrusion-cooked waxy-maize starch." *J. Cereal Sci.* 18: 277–286.

Politz, M. L., J. D. Timpa, and B. P. Wasserman. 1994a. "Quantitative measurement of extrusion-induced starch fragmentation products in maize flour using nonaqueous automated gel-permeation chromatography." *Cereal Chem.* 71: 532–536.

Politz, M. L., J. D. Timpa, A. R. White, and B. P. Wasserman. 1994b. "Non-aqueous gel permeation chromatography of wheat starch in dimethylacetamide (DMAC) and LiCl: extrusion-induced fragmentation." *Carbohydrate Polymers.* 24: 91–99.

Prudêncio-Ferreira, S. and J. A. S. Arêas, 1993. "Protein-protein interactions in the extrusion of soya at various temperatures and moisture contents." *J. Food Sci.* 58: 378–381, 384.

Qu, D. and S. S. Wang. 1994. "Kinetics of the formations of gelatinized and melted starch at extrusion cooking conditions." *Starch/Stärke.* 46: 225–229.

Queguiner, C., E. Dumay, C. Salou-Cavalier, and J. C. Cheftel, 1992. "Microcoagulation of a whey protein isolate by extrusion cooking at acid pH." *J. Food Sci.* 57: 610–616.

Ralet, M.-C., G. Della Valle, and J.-F. Thibault. 1991. "Solubilization of sugar-beet pulp cell wall polysaccharides by extrusion cooking." *Lebensm.-Wiss. u.-Technol.* 24: 107–112.

Ralet, M.-C., G. Della Valle, and J.-F. Thibault. 1993. "Raw and extruded fibre from pea hulls. Part I: composition and physico-chemical properties." *Carbohydrate Polymers.* 20: 17–23.

Riha, W. E. and C.-T. Ho. 1998. "Flavor generation during extrusion cooking." In: *Process-Induced Chemical Changes in Foods.* F. Shahidi, C.-T. Ho, and N. van Chuyen, eds. New York: Plenum Press, pp. 297–306.

Roussel, L., A. Vielle, I. Billet, and J. C. Cheftel. 1991. "Sequential heat gelatinization and enzymatic hydrolysis of corn starch in an extrusion reactor. Optimization for a maximum dextrose equivalent." *Lebensm.-Wiss. u.-Technol.* 24: 449–458.

Semwal, A. D., G. K. Sharma, and S. S. Arya. 1994. "Factors influencing lipid autoxidation in dehydrated precooked rice and Bengalgram dahl." *J. Food Sci. Technol.* 31: 293–297.

Tepal, J. A., R. Castellanos, A. Larios, and I. Tejada. 1994. "Detoxification of jack beans (*Canavalia ensiformis*): I—Extrusion and canavaline elimination." *J. Sci. Food Agric.* 66: 373–379.

Theander, O. and E. Westerlund. 1987. "Studies on chemical modifications in heat-processed starch and wheat flour." *Starch/Stärke.* 39: 88–93.

Ummadi, P., W. L. Chenoweth, and P. K. W. Ng. 1995a. "Changes in solubility and distribution of semolina proteins due to extrusion processing." *Cereal Chem.* 72: 564–567.

Ummadi, P., W. L. Chenoweth, and M. A. Uebersax. 1995b. "The influence of extrusion processing on iron dialyzability, phytates and tannins in legumes." *J. Food Process. Preserv.* 19: 119–131.

Unlu, E. and J. F. Faller. 1998. "Formation of resistant starch by a twin-screw extruder." *Cereal Chem.* 75: 346–350.

Van den Hout, R., J. Jonkers, T. Van Vliet, D. J. Zuilichem, and K. Van 'T Riet. 1998. "Influence of extrusion shear forces on the inactivation of trypsin inhibitors in soy flour." *Trans. Instit. Chem. Engin.* 76: 155–161.

van Zuilichem, D. J., G. J. van Roekel, W. Stolp, and K. van't Riet. 1990. "Modelling of the enzymatic conversion of cracked corn by twin-screw extrusion cooking." *J. Food Engin.* 12: 13–28.

Wang, W.-M., C. F. Klopfenstein, and J. G. Ponte. 1993. "Effects of twin-screw extrusion on the physical properties of dietary fiber and other components of whole wheat and wheat bran and on the baking quality of the wheat bran." *Cereal Chem.* 70: 707–711.

Wang, W. M. and C. F. Klopfenstein. 1993. "Effect of twin-screw extrusion on the nutritional quality of wheat, barley, and oats." *Cereal Chem.* 70: 712–715.

Whalen, P. J., M. L. Bason, R. I. Booth, C. E. Walker, and P. J. Williams. 1997. "Measurement of extrusion effects by viscosity profile using the Rapid Viscoanalyzer." *Cereal Foods World.* 42: 469–475.

Zhang, Y., C. M. Parsons, K. E. Weingartner, and W. B. Wijeratne, 1993. "Effects of extrusion and expelling on the nutritional quality of conventional and Kunitz trypsin inhibitor-free soybeans." *Poultry Sci.* 72: 2299–2308.

Zhao, J. and M. E. Camire. 1995. "Destruction of potato peel trypsin inhibitor by peeling and extrusion cooking." *J. Food Qual.* 18: 61–67.

Zhao, J. and M. E. Camire. 1994. "Glycoalkaloid content and *in vitro* solubility of extruded potato peels." *J. Agric. Food Chem.* 42: 2,570–2,573.

CHAPTER 8

Practical Considerations in Extrusion Processing

MIAN N. RIAZ

WHICH EXTRUDER TO PURCHASE

Food and feed manufacturers who are already using extrusion technology can answer this question very easily. They know the type of product and volume accepted by their market as well as any advantages their competitors may have in product appearance or functionality due to machinery differences. However, for new companies or even experienced food manufacturers who are considering entry into a rapidly growing food and feed market, the decision is not easy. They have a harder time selecting the components and configurations needed. But, when the homework is done, the information gathered, and the various factors considered, the number of practical alternatives is refreshingly small. Often the final decision is based on local factors such as cost and expected service from the extruder manufacturer. The options are numerous enough to guarantee confusion among first-time buyers. The fact that many extruder manufacturers are able to exist competitively indicates that most have found unique niches for their machines.

Following are some general considerations that should be contemplated when selecting an extruder:

- type of product(s) to be made
- type of raw material to be used
- size of extrusion plant that can be sustained with the expected market
- capital availability, investment cost, and recovery period
- processing rate, extruders do not run well when starved
- operating cost
- flexibility is costly
- hardened wearing parts for sustained usage—screw tips, barrel sleeves, die inserts
- relative energy requirements of machines considered, and local costs of steam, electricity, and possibly other types of mechanical energy
- local backup expectations and capabilities of manufacturer, availability of skilled operating and maintenance personnel, and expected needs for advice and field service from the equipment manufacturer
- used extruders/plants are available

In time, the extruder operators will become familiar with the capabilities of the extruder, its mechanical characteristics, and its maintenance requirements. In turn, this will generate increased confidence among the company's engineers, operators, and mechanics, and possibly lead to innovations that improve extruder efficiency. If experienced personnel are available, purchase of used equipment may be effective. If the foods/feed processor is already familiar with the abrasive properties of the product, initial specification of hardened wearing surfaces, including dies, may be appropriate. Automated controls also may be desirable, and their use may be considered for operations upstream and downstream from the extruder. It is wise to reconfirm your ideas through your personal information network, and possibly pilot plant trials, before large capital commitments are made.

Overflexibility and overcapacity can be expensive burdens. If the new product line does well, many extra features may never be fully used. Most extruders will not operate as intended over a large range of throughput and should not be run at greatly reduced rates. Extruders may be run full-out for only part of the day, while the packaging line runs continuously. But, this requires conveyors and additional bins for holding product that may not store well in bulk. Additionally, rework will be generated each time the extruder is started or stopped. It is best to assemble an extrusion plant with the various components as best balanced for throughput as possible. If the need for expanding capacity can be foreseen, available capital is better invested in larger tanks and expanded floor space.

WHEN TO BUY A TWIN-SCREW EXTRUDER

Single-screw extruders have successfully served the food industry over the last sixty years. With increasing consumer demands and new ideas, extruder manufacturers started working on twin-screw extruders twenty years ago because of flexibility and other advantages. Twin-screw extruders cost almost two times more than single-screw extruders, but they offer several advantages and flexibility that a single-screw extruder cannot offer.

The decision to buy a twin-screw extruder instead of a single-screw extruder depends on several different factors. These factors will help to make the decision to buy a single- or twin-screw extruder. If the product or process has the following characteristics, then a twin-screw extruder is the right choice.

- frequent product changeover
- products with high fat contents (above 17%)
- addition of a high level of fresh meat in the product (above 35%)
- uniform size and shape
- ultra-small product sizes (less than 1.5 mm)
- products made with low-density powder
- special formulations

COMMON EXTRUSION PROBLEMS AND THEIR SOLUTIONS

Sometimes during processing, an extruded product does not comply with the qualifications set for a product. The following are the most commonly observed problems in an extrusion system.

- production rates are too low
- wedging of product
- too much surging
- bulk density is too high
- bulk density is too low
- distorted product
- too many fines during processing
- product color is too dark
- product color is too light
- product is broken immediately after extruding
- moisture variation is too wide within the product

- moisture distribution is uneven
- extruder parts wear too quickly
- palatability is low

Everything happens for a reason, and troubleshooting is simply a matter of identifying a cause and then fixing it. The first step to take when identifying a problem is to gather accurate information about the raw material, hard wear, processing conditions, and final product.

A problem occurs for one of the following reasons:

(1) The problem was always there but was never identified as a problem.
(2) Something has changed gradually with time, and no one paid any attention until it became a major problem.
(3) Something happened suddenly.

There is a list of checkpoints that can help solve most of the above-mentioned problems.

CHECKLIST

- check the water supply to the preconditioner and the extruder barrel
- check the steam supply to the preconditioner and the extruder barrel
- check the indirect heating/cooling system of the extruder barrel
- check all parts for wear, especially screws and shearlocks
- check the shaft for wear or dent
- confirm the right configuration for the specific product
- calibrate the dry feed rate on a regular basis
- confirm your processing parameters
- check for die wear and be sure all the holes are open
- check knife for sharpness and its clearance against the die

Some of the common problems are surging, wedging, lower feed rate, product density, product breakage, and product color variation. Following are some helpful hints from Mair (1998), with modification regarding these problems.

SURGING

"*Extruder running empty then full just before the die. Products produced by surging are either low or high in density.*"

This is a common problem when an extruder is not running at full capacity. In this case, slow the extruder shaft rpm, and increase the feed rate, or plug some die open area. With time, extruder wear increases which results in backflow of the material, and this has a direct effect on surging. All worn parts should be replaced in order to reduce surging.

Surging is a major problem when a recipe is high in fat or moisture because of slippage and backflow. In this case, there are two options, reduce the oil or moisture content and/or reconfigure the extruder with more aggressive and positive conveying screws. Too fine grind is always a problem for conveying. Surging will occur if the material is too finely ground, because the material does not want to convey. To solve this problem, the grind needs to be a little coarser so that it will flow better. Remember, material that is too coarse is harder to cook, and a product with coarser particles is more susceptible to breakage.

Extruder screws usually have a different taper or leading and trailing edges affecting conveying and surging. Screws need to be indexed so the flights do not interfere, causing a problem like surging.

WEDGING

"The product flows unevenly from the die, being cut thicker at one side than the other."

One possible cause for this problem is that the product is being held back at one side by knives as each knife blade travels over the die. This problem can be fixed by reducing the knife blade thickness or increasing the knife speed by reducing the number of knives. Adjusting the blade angle or grind will also help. Make sure the knife does not lie flat against the die face.

Another possible reason for this problem may be a product that flows unevenly from the die, being cut thicker at one side than the other. In a single-screw extruder, velocity at the outside of the die tends to be higher than velocity at the center. This radial velocity gradient can be reduced by inserting a screen, inserting a backup (primary) die system before the final die, or creating a dead space between the end of the screw and the die. Another way to handle this problem is to increase the feed rate, or increase the extrudate viscosity by lowering oil or water addition. Product flow is affected by die design. Increasing the radius of tight corners will increase flow, and shortening the land length on the side with the slowest flow will also help.

LOWER FEED RATE

"Suddenly feed rate needs to be adjusted because the extruder is filling up."

This situation usually occurs when screw parts are worn and causes backflow of the material. The extruder fills up at a fixed die pressure, reaching the point at which extrudate backs up to the inlet. The feed rate needs to be adjusted until worn parts are replaced. This situation can be corrected temporarily by increasing extruder rpm. Cooling the extruder barrel at the feeding zone produces positive results.

To make a decision to replace the worn parts, always measure the clearance between the screw and liner. When this clearance doubles, it is time to replace the worn parts. It also depends upon the product that is being extruded. Some products have less effects than others.

Another possible reason for sudden lower feed rate may be the increase in viscosity of the extrudate, which causes backflow. This problem can be solved by reducing the oil or water addition in the recipe. Some time variation in the raw material, like fresh meats, can increase the viscosity by having more fat or water, which ultimately contributes in lowering the production rate.

Sometimes steam injection is too close to the feeding section and steam flow causes a feedback problem. Be sure that steam is being injected at the proper place and is being distributed equally in the barrel. Sometimes steam can act as a "shearlock" and prevent dry feed from being conveyed down to the extruder. Always clean the screw before starting an extruder. Sometimes old feed remains in the liners and can reduce the feed rate.

A misaligned or bent extruder shaft can cause a "shearlock effect," causing feed to back up. Sometimes the screws configuration can be backward. Be sure that all the screws are in the correct configuration and none have been put in backwards. Also, check all the shearlocks for correct configuration and angles. The recommendation is to start with the smallest diameter close to the feed inlet. Be sure the grind is the same and the source of ingredients has not changed. Sometimes newly harvested ingredients grind differently than older ingredients.

VARIATION IN PRODUCT DENSITY

"Suddenly product density is too low or too high."
Density is a function of expansion, size, and shape.

This situation could occur for several reasons: dies are worn out, grind is coarse, oil content of dry feed is too high, feed rate has changed. This problem can be solved by checking worn dies, increasing feed rate, checking for worn screw and liner, increasing the conditioning temperature, increasing or decreasing the water in the preconditioner, adjusting the feed rate, increasing the steam injection into the extruder barrel, cooling the barrel, increasing the starch content in the recipe. Always check the product specifications for size and shape. A minor change in specification can have a major effect on the product density. If dies are worn, the extruder operator may be controlling size by reducing expansion ratios, thus increasing bulk density.

PRODUCT BREAKAGE/DISTORTION

"A product can get distorted if it is too soft, or is undercooked. Sometimes products can be damaged by knives and conveying systems."

This problem can be solved by changing the following parameters:

(1) Make sure the knife speed is not too high
(2) Increase the number of blades
(3) Check for plugged dies and increase the number of holes
(4) Check knife hoods for proper installation, sometimes product can hit the hood and break
(5) Check the product for excessive oil or water
(6) Reduce air velocity for a negative air system
(7) Reduce knife width and angle of attack of knife blades
(8) Check for small unground pieces of ingredients in the final product that may act as breaking points

COLOR VARIATION

"Suddenly products vary in color."

Color is usually a function of degree of cook, particle size of the ingredients, and added color.

Coarsely ground ingredients make products look darker. Specification and amount of color added also affect the final product color. Color variation can be controlled by changing the following parameters:

(1) Check grind, and be sure the size has not changed
(2) Check variation of the rate of dry feed and addition of color
(3) Blend the color evenly in the premix
(4) Check feed formulation changes
(5) Note that natural colors can vary more than artificial colors
(6) Note that some natural colors vary in their intensity, i.e., iron oxide

EXTRUDER WEAR

A typical extruder can run anywhere between 200 to 20,000 hours before parts are worn enough to be replaced. The life of the extruder screw depends upon the raw material to be extruded and several other factors, like moisture, conditioning, configuration, fat content, metallurgy, and misalignment.

Raw Material

Different ingredients have different effects on the wear of an extruder. For instance, sugar and salt will have more effects on wear than other ingredients. Also, the coarse grind of ingredients will have more effects on wear than the fine grind.

Moisture

Higher moisture content decreases extruder wear. Direct expanded products with moisture of 12–16% can wear the parts in 350–450 hours. Products with moisture of more than 25% can increase the life of an extruder screw more than 10,000 hours.

Preconditioning

Preconditioning can more than double increase the life of an extruder screw. More detailed information about preconditioners and their advantages are discussed in Chapter 6, "Preconditioning."

Configuration

Intensive use of shearlocks and reverse flight in a configuration will reduce the free volume and will have an adverse effect on extruder wear.

Oil Content

Oil content in a recipe works as a lubricant and reduces the wear of the screw. Therefore, a high amount of oil content will increase the screw life.

Metallurgy

Extruder parts made from extremely hard alloys will have more life than parts made from a cheap source.

Misalignment

Metal to metal is the major cause for the extruder wear. This will cause scour marks and sharp edges on screw flights. If you still have knocking noises after the adjustment of a misaligned machine, look for nonparallel screw flights or material trapped between screws. If you still have the problem, suspect a bent shaft or bearing housing problem.

START-UP SEQUENCE FOR A TYPICAL EXTRUDER

Each extruder has its own start-up sequence, which helps to stabilize and ensure the quality of the extrusion process. Most extruder manufacturers supply guidelines to start up their equipment. An extruder operator must follow these guidelines in order to maximize the life of the extruder parts. There are several prestart and start checkpoints which can help to quickly stabilize an extruder and ultimately give more production time.

PRE-START-UP CHECKS

There are several pre-start-up checks that can help an operator quickly stabilize an extrusion process. These pre-start-up checks include equipment checks and utility checks.

Equipment Check

Before start-up can occur, it is necessary to assemble the extruder. The procedure for assembling an extruder will depend upon the manufacturing style, model, and specifications. As discussed in this chapter, there are several different types of extruders available. Each extruder will require a special kind of assembly setup. Assembling a single-screw extruder as compared to a twin-screw extruder is less complicated and relatively easy. Whereas, with a twin-screw, a lot of attention needs to be paid to the screw configurations. Similarly, an extruder that has a one-piece screw and barrel is much easier to assemble compared to an extruder that has several parts. As a rule of thumb, all parts for an extruder should be laid out systematically for a quick and orderly fashioned assembly. All parts should be checked visually during assembly for defective or excessive wear.

The following are general points that can be considered for most extruder assembly.

- Check bolts that secure the inlet head to the yoke. Make sure all bolts are secure and properly in place.
- Check bolts that fasten the extruder barrel clam-shell together to make sure all bolts are in place and are properly secured.
- Check head supports to make sure head support pins are properly in place and head supports are properly adjusted.

- Check bolts that fasten the die to the discharge of the extruder to make sure they are installed and properly secured.

Utilities Check

Electrical Supply

- Check the main power supply switches and the main DC drive switch to make sure they are on.

Steam Supply

- Check preconditioner steam injection to make sure that supply valves are open and required steam pressure is on. Open the injection valve to flush condensate from supply lines.
- Check barrel steam injection to make sure that supply valves are open and required steam pressure is on. Open the injection valves to flush condensate from supply lines.

Water Supply

- Check water supply to pumps that supply water to preconditioner and extruder barrel injection to make sure supply valve is open.
- Check cooling water supply to hot oil heat exchanger (only if barrel is heated by oil) to make sure supply valve is open.
- Check cooling water supply to extruder barrel sections to make sure supply valve is open.

Air Supply

- Check the main air supply valve if air motors are being used.

Thermal Fluid System

- Check the thermal fluid system to make sure supply and return lines are connected to the proper zones.

Instrumentation and Electronics

- Check thermocouples to make sure that they are installed and that leads are connected to the proper zones.

PRE-START-UP SEQUENCE

The purpose of the start-up procedure is to bring the extruder to operating and stable condition as quickly as possible. In some cases, raw material is very expensive, and by delaying the start-up, we can lose raw material. In most cases, under- or overprocessed material can be mixed back into the raw material. But in some cases it is not possible, and this will increase the cost of the operation.

Before we can start feeding the ingredients to the extruder, the system should be brought as closely as possible to stable operating conditions. The following points can help to insure a quick and steady start-up of an extruder.

- Fifteen minutes prior to start-up, set the zone control temperatures to the desired setting and start the heating units.
- Extruder barrel steam injection may be used to assist in preheating the extruder barrel to the proper temperature.
- Two minutes prior to start-up, start the lubrication pump.

START-UP SEQUENCE

- Check the diverter spout. Make sure it is set to bypass product from the extruder barrel.
- Start the preconditioner and set it to approximately medium speed.
- Start the feeder. Set it to a low rate.
- Wait till product begins to flow down the downspout, then adjust the desired steam flow to the preconditioner.
- Start the water pump that delivers water to the preconditioner and adjust it to the set point.
- Check the consistency of product coming out of the cylinder when the flour is warm and moist or at the desired consistency.
- Start the water pump which delivers water to the extruder barrel and adjust it to a set point.
- Immediately start the extruder and turn the rpm slowly.
- Divert the material into the extruder barrel.
- Immediately, over a three- to five-second time span, turn the extruder up to the desired speed.
- Watch very closely for motor overload and for excessive pressure in the cone section, if either of these are observed, one of the following has most likely occurred:

a. Excessive feed passed into the extruder barrel. Correct this by passing feed intermittently and reducing the feeder rpm.
b. There is inadequate moisture in the extrudate. Adjust moisture levels by increasing water flow to the extruder barrel.
c. There is a blockage in the die. Correct this by diverting the material flow into the extruder immediately, and begin shutdown procedures.

- If overloading does not occur and the product is flowing out of the die, make the following checks:

a. Check the extrudate for moisture. If it is very wet, begin increasing the extruder rpms or reducing the amount of moisture to the extruder.
b. If steam injection is required, adjust the steam injection rate.
c. When the product appears to be of proper consistency to cut, start the knife drive.
d. Adjust the feeder, moisture, and temperature settings to achieve your desired product.

EXTRUSION "RULES OF THUMB"

Following are the rule of thumb from Lusas and Riaz (1999) with modifications.

PRODUCT FORMULATION

Limitations do not permit a review of optimization of formulas for specific nutritional needs or least costs. Manufacturers should retain the services of nutritionists and formulation specialists as needed. It must be remembered that some type of cohesive network is needed to keep the formed chunks together. For the most part, this is accomplished by starch, but functional proteins may be involved—especially in the production of texturized soy protein products. The formula must contain enough cereal grains and their starchy fractions for the dextrinized/gelatinized starch to cement product pieces together. Meat and fish meals and corn gluten do not contribute to cohesiveness. Oil and fat generally are detrimental to the puffing of starch and the texturization (striation) of soy protein at levels over 6%. If obtaining a porous and expanded product is not important, as much as 15–18% fat can be included in products made on single-screw extruders, and 20–22% fat can be included in products made on twin-screw extruders.

In addition to the interference of fat in product expansion, natural and added antioxidants are partially lost by volatilization during flash-off of expanding products as they exit the die (Fapojuwo and Maga,

1979). Vitamins are often destroyed during extrusion processing—despite efforts to protect them by encapsulation (Lee et al., 1978; Maga and Sizer, 1978). Lactose in dry milk and whey products brown during moist heat processing (Raricot et al., 1981). The preferred practice is to include minerals and heat-stable color materials in the extruder feedstock mix to ensure uniform distribution but to hold back the fat, antioxidants, vitamins, milk/whey ingredients, and flavoring materials for enrobing onto the formed product after extrusion and drying.

SIZE REDUCTION OF INGREDIENTS

Some extruders are capable of grinding whole corn and soybeans, but many people feel that these operations can be accomplished more efficiently with dedicated equipment like roller or hammer mills. A general rule of thumb is that the extruder feedstock should not contain any particles larger in size than 1/3 the diameter of the die holes. Passing the ingredients or extruder feedstock mixture through a positive classification device like a powered sieve or screen, or a hammer mill equipped with a suitable-sized screen, is recommended before extrusion. Equipment also should be protected by magnets and metal detectors to avoid damage from tramp metal.

MIXING

Various mixers are available for the dry feedstock. The minimum acceptable mixing type can be determined by stopping the mixer and taking samples randomly from the contents. The samples are next analyzed for an ingredient (like salt or a mineral component). Minimum acceptable mixing is the time required to reach the lowest stable standard deviation of analysis of the sample groups. The reader is referred to a statistics handbook for further instructions. Unfortunately, ingredients may "unmix" under certain conditions. For example, heavy mineral components may settle to the bottom of the trough of a slowly moving conveyor. Dropping the mixture through a conveyor spout into a cone-formed pile may result in the heavier particles settling on the outside of the cone, especially at the bottom of the cone. In some instances, materials may be wetted and worked into doughs before being fed to the extruder.

PRECONDITIONING

During gelatinization, the starch granule absorbs water, swells, and loses its crystallinity, while in dextrinization, which is favored by extrusion at lower moisture contents, the starch granule is torn apart physically. Both processes cause the starch to become more readily digested; however, dextrinized corn flour has greater cold water solubility. Where maximization of gelatinization is desired, time is required for the added moisture to equilibrate with the cereal fraction before extrusion. Wetting of finely ground cereals can be difficult if the dry flour is added directly at the extruder feed inlet with an accompanying stream of water. It has become common to premix and preheat feed stocks in a preconditioner prior to introduction into the extruder. Preconditioners are less expensive to build and have retention times up to three times that of the extruder. However, care must be taken to avoid buildup of gluten on the preconditioner paddles when processing wheat flour.

Pressurized preconditioners can precook the cereal feedstock before introduction to the extruder. Most manufacturers of cooking extruders now offer steam preconditioners (see Chapter 6). Similar precooking units are sometimes installed before pellet mills.

Heated noncondensable gases (entrained air) and excess steam can cause serious surging problems when cooking and forming are conducted in the same extruder barrel. Precooking in the conditioner reduces these problems because the feedstock is hot when introduced into the extruder at atmospheric pressure. Some manufacturers have achieved the same effect by installing a vent to depressurize cooked feedstock in the extruder barrel entering the final forming section. In some operations, a vacuum is pulled on the vent to increase evaporation of moisture and remove volatile compounds. Also, additional heat-sensitive ingredients can be added downstream into the cooked feedstock.

FEEDING THE EXTRUDER

The difficulty of wetting dry powders to enhance feedstock traction (picked up by the screw) at the inlet of the extruder has been mentioned. This is less of a problem in twin-screw extruders than in single-screw machines, but still can be significant. Preconditioning enhances screw pickup by wetting the product. Problems exist in picking up doughs and sticky products and can be reduced by building a sloped feed throat or undercut feed throat hopper in the feed end barrel that keeps the dough from riding on top of the revolving screw; building a packer or second short, powered parallel screw into the feed throat; building a multiscrew live-bottom feeder at the inlet; or using a square-throat positive displacement pump to feed the extruder, provided the feedstock is pumpable.

Generally, a plug of product must be maintained between the feed and the cooking sections to prevent blowback of steam. This can be done by cooling the transition section with a cold water jacket often in conjunction with using an appropriately sized shearlock.

WHERE TO ADD INGREDIENTS IN THE EXTRUDER

All the ingredients have an appropriate place for addition into the extrusion system. Ingredients not introduced at the proper place may have an adverse effect on the final product and may result in poor and inconsistent quality.

Recommended Places for Addition

Following are the recommended places to add solid and liquid ingredients.

Dry ingredients are usually metered into the preconditioner from a live bin or by a feeding screw. If no preconditioner is being used, then these dry ingredients can be directly metered into the extruder barrel.

Liquid fat can be mixed with the dry ingredients in a mixer. Other options include metering directly to the preconditioner or extruder barrel (only twin-screw) depending upon the product and desired specification. Liquid fat can also be blended with fresh meat in premixing, if meat is part of the recipe.

Emulsified fresh meats can be metered into the preconditioner or sometimes directly into the barrel of a twin-screw extruder only as a part of the dry feed rate.

Process water should be metered into the preconditioner or extruder barrel by using manual flow meters and valves.

Process steam should be injected directly into the preconditioner or extruder barrel using a flow meter and valves.

Oil or other liquid coating can be metered into a coating machine and controlled by a product weighing device. These can also be added to a product using a batch enrobing mixer.

Powder for enrobing should be metered into a continuous coating machine from a powder feeder and controlled by a product weighing device. This can also be added to the product using a batch enrobing mixer.

COOKING THE PRODUCT

Optimum screw and barrel designs vary, depending on the sequence in which the product is to be heated, sheared, and texturized (especially in the case of proteins). This topic is too lengthy to cover in this chapter.

TEMPERATURE AND PRESSURE CONTROL AT DISCHARGE

Temperature control, which in turn affects the extent of flashing and expansion after the die, is important when products of specific densities must be maintained. As mentioned earlier, expansion and firmness of extruded products depend primarily on the amount of starch, functional protein, and fat present in the formula. Cooling of the product before it exits through the die reduces expansion and helps maintain integrity of fragile structures. Most proteins are heat denatured before they get to the die plate, but soy and other oilseed globulins will align to form elongated steam cells if allowed to flow unagitated after processing through a pipe or other suitable configuration before discharge and cutting. The resulting multilaminate pieces have the appearance of striated meat.

DRYING AND COOLING

Moisture is reduced by evaporation, which in turn cools the extruded product. An approximately 4–7% reduction of moisture is obtained by flash-off as hot products expand at the die. Additional drying can be done atmospherically if sufficient time and low humidity air are available. If needed, drying is typically completed by supplemental heat in a continuous band-type oven.

COATING AND ENROBING

Hydrolyzates and water-based emulsions can be gently mixed with the extruded pieces before going to the dryer. A variety of equipment is used for coating-enrobing, including horizontal reels, ribbon mixers, and open-ended barrel-type tumblers equipped with vanes to gently tumble the product. Fat, heat-sensitive antioxidants, vitamins, and flavorings are typically sprayed or dusted onto the dried cooled product as the final processing step. These materials are expensive and may be moistened to ensure adhesiveness to the product to minimize their loss as fines. In pilot plant or small-scale operations, enrobing can be done batch-wise in a slowly turning horizontal ribbon blender or inclined barrel tumbler. If sugar-glazing is done, as in the case of presweetened ready-to-eat cereals, the enrobed product may be returned to a second series of ovens and coolers for drying.

PACKAGING

If the product is to be sold in packages, the packaging material should include a fat barrier to prevent wicking and development of unsightly spots on the exterior.

CLEANING OF THE EXTRUDER AND SCREW

Provisions differ for cleaning the extruder screw. In most segmented-barrel single-screw extruders, the barrel sections unbolt and pull off piece by piece. However, occasional small models are built with a clamshell design that enables splitting and swinging open the two halves of the barrel. The smaller twin-screw extruders are built with either a clamshell barrel consisting of segments that are bolted together or solid barrel sections that are pulled off one at a time. A twin-screw extruder manufactured in France includes a railing for pulling the entire barrel assembly off the screws. Pulling the individual worm segments from the shaft may require considerable effort on either single- or twin-screw machines because of product burn-on. It is generally best to organize operations that allow sustained extrusion runs, especially when producing animal foods. However, the ease of disassembly/reassembly and cleaning may be a significant factor in selecting extruders for specific operations.

REFERENCES

Fapojuwo, O. M. O. and J. A. Maga. 1979. "Butylated hydroxyanisole (BHA) retention during the extrusion of corn." I. Agr. Food Chem. 27: 822–824.

Lee, T., T. Chen, G. Alid, and C. O. Chichester. 1978. "Stability of vitamin A and provitamin A (carotenoids) in extrusion cooking processing." AIChE Symposium Series Food, Pharmaceutical and Bioengineering—1976/77. 74: 192–195.

Lusas, E. W. and Riaz, M. N. 1999. Introduction to extrusion and extrusion principles. In "Feeds and Pet Food Extrusion Manual." Ed; Riaz, M. M. Food Protein R&D Center, Texas A&M University, College Station, TX.

Maga, J. A. and C. E. Sizer. 1978. "Ascorbic acid and thiamine retention during extrusion of potato flakes." Lebensm - Wis. u. -TechnoL. 11: 192–194.

Mair, C. 1998. "Solving common processing problems." Paper Presented at Petfood Forum 98. Chicago, IL, Mar. 30–Apr. 1.

Raricot, W. F., L. D. Satterlee, and M. A. Hanna. 1981. "Interaction of lactose and sucrose with cornmeal proteins during extrusion." I. Food Sci. 46: 1,500–1,506.

CHAPTER 9

Extruders in the Food Industry

ERIC SEVATSON
GORDON R. HUBER

INTRODUCTION

DISCUSSED in this chapter will be high-temperature/short-time (HT/ST) single- and twin-screw cooking extruders used in the food industry. This will not be an extremely technical discussion because most of the intimate details regarding the science of extruders are covered in other chapters in this book. Covered in this chapter will be the more common uses of cooking extruders in the food industry with an outline and description of the process flow for each of these processes including raw materials used and their specifications.

HT/ST single- and twin-screw cooking extruders should not be confused with cold forming extruders, which are also referred to as cold formers or, simply, formers, like those used in the pasta industry. Shown below is a diagram listing all of the common components of single- or twin-screw extruders followed by a chart outlining the functions of those extruders (Figure 1).

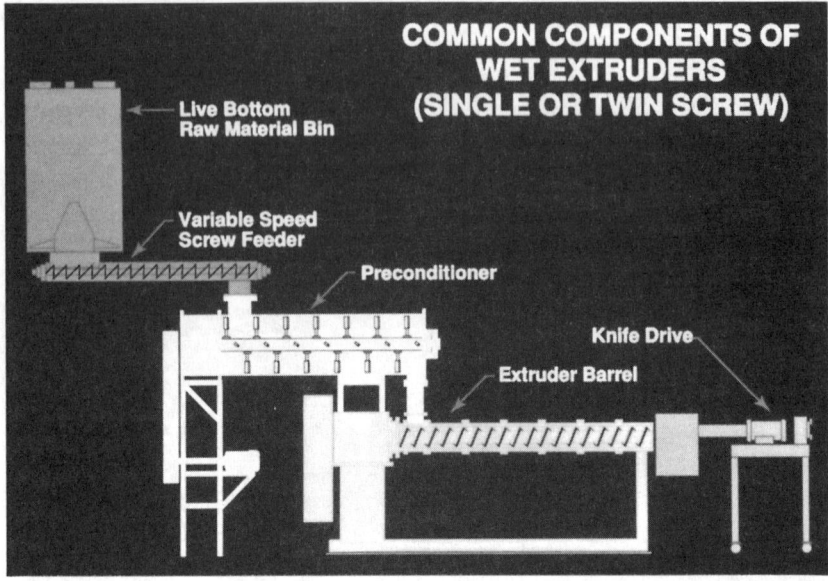

Figure 1 Common components of wet extruders—single- or twin-screw.

Functions of a Single- or Twin-Screw Extruder.

Feeding	Phase Transition
Conveying	Diffusion
Mixing	Heating/Cooling
Distributive	Pumping
Dispersive	Retention Time
Hydration	Extruding
Kneading	Forming
Sealing	Venting
Compression	Expansion

HISTORY AND USES OF EXTRUDERS IN THE FOOD INDUSTRY

HISTORY

The foundation of food extrusion lies in the snack food and ready-to-eat breakfast cereal industries. History shows that the first application of the use of a cooking extruder in the food industry was in producing an expanded cornmeal-based snack in the mid-1940s on a single-screw extruder. Most likely, this was either an expanded yellow cornmeal ball or curl that was then coated with some type of seasoning such as cheese and salt. This is very similar to some of the products still offered today by many snack producers.

Currently, there are more single- than twin-screw extruders in use in the food industry. The main reason for this is that they have been around longer than twin-screw extruders. As mentioned above, single-screw extruders were first used in the 1940s, while twin-screw extruders were not developed for the food industry until the early 1980s. However, twin-screw extruders are rapidly becoming the extruders of choice in the food industry. Why is this? I heard one "old-timer" in the food industry say, "a twin can do everything a single can, but a single can't do everything a twin can." Basically, this quote relates to the fact that twin-screw extruders have more versatility than single-screw extruders. They have a greater affinity to overcome year-to-year and crop-to-crop differences in raw materials, are better suited to extrude complex formulas with "difficult" ingredients, have a better conveying/pumping ability, and have better heat transfer.

When I first became involved with food extrusion in the mid-1970s, there were probably only four suppliers of food extruders, and at that time, those were only single-screw extruders. Most of those extruders were used to produce snacks, ready-to-eat breakfast cereals, and textured vegetable proteins. By the late 1990s, there were at least ten suppliers of food extruders.

Suppliers of Food Extruders

- Wenger, U.S.
- Buhler, Switzerland
- Pavan Mapimianti, Italy
- Krupp Werner & Pfleiderer, Germany
- Clextral, France
- APV Baker, England
- Maddox Metal Works, U.S.
- Extru-Tech, U.S.
- American Extrusion, U.S.
- Lalasse, Holland

The more common uses of extruders in the food industry will be discussed here. Other less prevalent uses of extruders by this industry are in the production of baby foods, various types of grain/legume analogs, bran stabilization, precooked or thermally modified starches, flours, and grains, beer powders, beverages, cheese and casein, food gums, reformed fruit bits and sheets, topping and bakery analogs, coextruded products, precooked pasta, breadings, and bread-like products.

COMMON USES OF SINGLE- AND TWIN-SCREW COOKING EXTRUDERS IN THE FOOD INDUSTRY

The production of textured vegetable proteins, ready-to-eat breakfast cereals, and direct expanded (DX) and third generation (3G) snacks require processing with very simple single-screw extruders to more complex twin-screw extruders. Direct expanded (DX) corn snacks require only a short barrel single-screw extruder with no preconditioning and only use a small amount of additional water injected into the barrel during processing (Figure 2). Extruders for the production of third generation snack pellets, sometimes also referred to as half products, use a long (25.5:1 barrel length-to-bore diameter, commonly expressed as L/D) barrel twin-screw extruder with separate cooking and forming sections, vacuum vent, and preconditioning and use steam and water injection into the extruder barrel and the preconditioner (Figure 3). However, currently, research is being done by some manufacturers to

Figure 2 Single-screw snack extruder with snack meal feeder (upper right)—no preconditioner, short barrel.

Figure 3 Third generation snack extruder with 25.5:1 L/D twin screw (right to left), raw material supply bin, feed screw, preconditioner, and barrel vacuum vent.

try and produce 3G snacks on short (7.5:1 L/D) twin-screw extruders with preconditioning but no vacuum vent.

TEXTURED VEGETABLE PROTEIN (TVP) PRODUCTION

TYPES OF EXTRUSION-COOKED TEXTURED VEGETABLE PROTEINS

Meat extenders represent the largest portion of textured vegetable food proteins. These products are hydrated to 50–65% moisture and blended with meat or meat emulsions to extend these food products to levels of 20–30% or higher.

Simulated meat analogs, which are also referred to as just meat analogs, use extrusion cookers to transform vegetable protein sources directly into varieties of simulated meat analogs that are consumed as is and have the appearance, texture, and mouthfeel of meats. "Extrusion technology has developed to the point where textured vegetable proteins can be formed into a fibrous matrix almost indistinguishable from meat" (Central Soya, 1990).

Textured vegetable proteins can be added to meat as extenders or can be consumed directly as simulated meat analogs. "Breaded chicken patties with as much as 30% of the meat replaced were actually preferred to all-meat patties by a majority of participants in taste tests conducted at the Indiana State Fair" (Central Soya, 1990). Meat analog pieces can be flavored and formed into sheets, disks, patties, strips, and other shapes. There are meat-free hot dogs, hamburgers, chicken patties/nuggets, hams, sausages, meat snacks, and loose meat products for chilies, tacos, and spaghetti available that are difficult to tell from the real thing. In the U.S., most uses of meat extenders are as additives or extenders to other food products. Consumption of TVP-type products is on the rise in the U.S. and worldwide (Figure 4).

In India, China, Taiwan, Japan, and South Korea, TVP is eaten directly as a flavored or seasoned piece usually as a side or main course portion of the meal. A good example of a completely meat-free meat analog is flavored bacon bits. Some of the consumption of TVP in parts of the world is based on religious, cultural, or economic reasons. An example of this includes the vegetarian diets of most Indians. TVP-type

(a)

Figure 4 Various types of TVP products.

(b)

(c)

Figure 4 Various types of TVP products.

(d)

Figure 4 Various types of TVP products.

products have been used by worldwide relief agencies to help feed famine plagued peoples in impoverished countries and are widely used in child school nutrition programs. Because of their low moisture and water activity, storage, shelf-life, and handling under poor conditions do not become problems. They can be vitamin and mineral fortified and make an ideal protein source. Because TVP is cholesterol free and can be processed as a low fat food (with less than 1% saturated fat), it comes with some positive health benefits. The difficulties and problems associated with processing and handling real meats are not found in the textured vegetable proteins industry. Ease of use is a plus for any type of TVP (Figure 5).

RAW MATERIALS USED FOR TEXTURED VEGETABLE PROTEINS

Defatted soy flour, flakes, and grits, soy concentrates and isolates, mechanically extracted soy meal, wheat gluten, and other legume/grain sources are raw materials that are used in producing textured vegetable proteins. Textured vegetable protein products have traditionally used soybean proteins as the protein source. Recently, we have seen the limited use of other plant proteins such as wheat gluten (protein), peas pro-

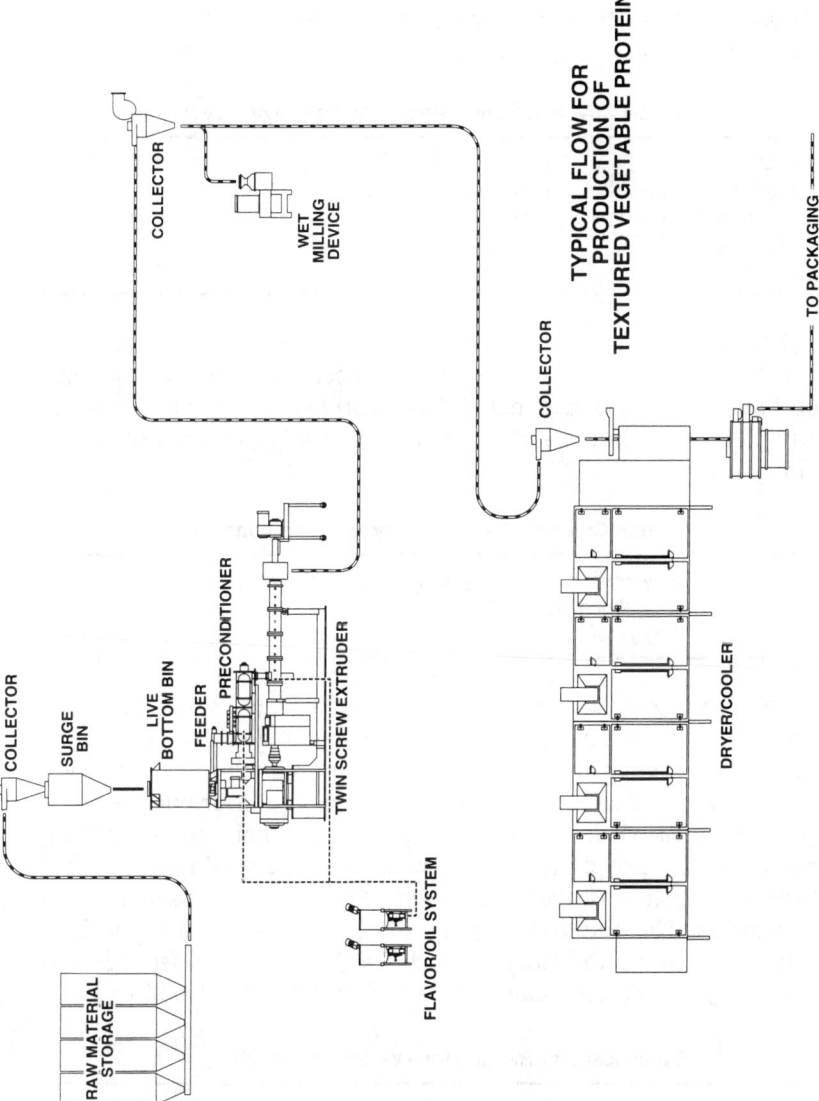

Figure 5 Typical flow for production of textured vegetable protein.

tein concentrates and isolates and defatted peanut flour, de-glanded cottonseed flour, canola (rapeseed) protein concentrates, and sesame, sunflower, and lentil (dal) proteins for this source. In some cases, recipe blends use a combination of these ingredients. Within limits, any plant protein source could be texturized.

Basic Raw Soybean Composition (Central Soya, 1990).

Protein	38.0%
Soluble carbohydrates (sucrose, stachyose, raffinose, other)	15.0%
Insoluble carbohydrates (dietary fiber)	15.0%
Oil (lecithin 0.5%)	18.0%
Moisture	14.0%

Soybean proteins are the single largest source for the manufacturing of TVP type products worldwide. This is because of the simple economic law of supply and demand. The worldwide soybean supply is plentiful, and overall, it is a cheap, relatively easily managed source of protein.

Protein Content of Various Soybean Fractions.

Soy flour, flakes, and grits	52–53% (as is)
Soy concentrates	66–69%
Soy isolates	88–92%

Defatted soy flour has been used most extensively as a meat extender due to its low cost. Meat extenders from defatted soy flour have the advantage of retaining meat juices and fat and reducing cooking loss. Soy grits and flakes are chemically the same as the flour but are physically different in particle size and require additional preconditioning prior to extrusion. Soy concentrates and isolates are used to make TVP products, but because of their higher cost, they are rarely used as complete formulas. The addition of concentrate or isolate in the soy flour formula can improve the final product. Concentrates and isolates can increase the water holding capabilities and protein content of TVP.

Specifications Ranges for TVP Raw Materials.

Protein Dispersibility Index (PDI)	50–70%
Protein*	52–92%
Percent fat	0.5–6.5%
Percent fiber	0.5–7.0%
Particle size	45–150 microns

*depending upon source, see table above

The Protein Dispersibility Index (PDI) measures the total protein that is dispersed in water under controlled conditions of extraction. The Nitrogen Solubility Index (NSI) is another method used to quantify the quality of the protein found in the vegetable protein source. However, PDI is now the preferred method because it is generally recognized as being more accurate and reliable. TVP products have been produced with raw materials having a PDI of 50 to 70. Normally, the higher the PDI number, the better the final product. In rare cases, some soy materials can have a PDI value that is too high.

Fat levels in the various raw materials used to produce TVP range from 0.5 to 6.5%. The most commonly used raw material, defatted soy flour, is 1% fat. As fat levels increase in the raw materials, the extruder needs to increase processing temperatures and shear energy inputs. This is the only way that product quality can be maintained.

Fiber levels can range from 0.5 to 7.0%. Fiber has a negative effect on TVP quality. It is understood that fiber partially blocks some of the cross-linking of the protein macromolecules which can affect structure and texture.

Particle size has a very wide range. Some raw materials can go as fine as 38 micron or as coarse as 180 micron. Particle size is important in maintaining proper textural properties in TVP. Very fine grind flours when wetted can cause clumping, while particles over 180 microns in size are difficult to premoisten. It is best to use raw materials with particle sizes between 150 micron and 45 micron.

MODIFICATIONS TAKING PLACE IN THE EXTRUDER DURING TVP PROCESSING

Proteins are effectively denatured during the moist thermal process of extrusion. Denaturation of protein "lowers solubility, renders it digestible and destroys the biological activity of enzymes and toxic proteins" (Smith, 1975).

As mechanical and thermal energy are applied during extrusion, the macromolecules in the protein portion of the raw material lose their native, organized structure ("memory") and form a continuous plastic-like mass. The extruder barrel, screws, and die take advantage of this situation and align these macromolecules in the direction of flow. This explains the laminar appearance of most TVP foods. As this is all taking place, there is a molecular cross-linking between proteins that reshapes the product and enhances the reforming of this matrix into its unique expandable structure.

Vegetable proteins have a raw, bitter, unpleasant flavor. Most of the flavors associated with these obnoxious tastes are volatile and, therefore, are eliminated through the extrusion process. These compounds are flashed off with the steam as the TVP is being preconditioned or as it enters atmospheric conditions at the extruder die. This reaction starts in the preconditioner where steam is being applied and continues through the extruder where additional steam is being added and the extruder's barrel where heat and frictional energy are developed by the extruder's screws.

During extrusion of TVP, a homogeneous, irreversible, bonded dispersion of all minor ingredients throughout a protein matrix is developed. This insures the uniformity of all added ingredients within this matrix. Also provided in this adaptation is the potential for minor ingredients to be closely associated with the cross-linking site or other desirable chemical and physical transformations.

At the end of the extruder, the TVP is shaped with the final die and then cut and sized into its final shape by a rotating cutting knife.

Textured Vegetable Protein Extrusion Operation

Single-screw extruders are still by far the extruder of choice for manufacturing TVP-type products. However, it has been found that sometimes twin-screw extruders are able to produce acceptable TVP with raw materials falling outside of the required ingredient specifications mentioned above. Single-screw extruders are usually limited to only using ingredients within these prescribed specifications. This goes back to the statement the "old-timer" made about one of the differences between single- and twin-screw extruders. Because of the twin-screw extruder's ability to cope with less constraining raw material specifications, they are quickly catching up to single-screw extruders in TVP production.

The particular configuration of the extrusion system is specific for TVP production. Preconditioning and extruder screw and die configuration are important factors in producing acceptable TVP on an extrusion system.

Not all extrusion processes require preconditioners, but it is particularly useful in the manufacturing of TVP. Steam and water are added to the preconditioner prior to being added to the extruder. Water addition in the preconditioner ensures that each individual particle is evenly premoistened and improves the stability and final product quality. Typically, a moisture content of 10–25% is desired in the preconditioner. The addition of steam in the preconditioner promotes moisture penetration and starts the cooking process. Before product enters the extruder, the product temperature can be raised to 65–100°C with the addition of steam in the preconditioner. Preconditioners are vented to avoid excess steam and to allow removal of volatile flavor components.

Coloring agents, heat stable flavors, and other liquid additives can be added into the preconditioner. This would permit even, thorough continuous mixing of all ingredients.

"Without the use of preconditioners, it is difficult to make good laminar structured TVP. Unpreconditioned vegetable proteins have a strong tendency to expand rather than laminate due to non-uniform moisture penetration which does not allow uniform alignment of protein molecules. Using a preconditioner can allow for the use of a raw material with a larger particle size" (Rokey, 1990).

The extruder screw profile or configuration is specifically designed to produce TVP. After hydration and heating, the protein molecules are unwoven. Heating and hydration were started in the preconditioner and continue in the extruder where steam and water can also be directly introduced as well as the heat generated by the frictional forces and external heating of the extruder barrel. "The extruder screws now take these single un-woven protein bodies and realigns them in a laminar, stretched and twisted condition into an appearance of a meat like structure" (Rokey, 1990). As retention time and the shearing effects increase as the vegetable proteins continue along within the extruder barrel, more cross-linking occurs between these protein stands. This causes an irreversible reaction with the proteins. The "memory" of these proteins has now been reset and modified by the extruder and cannot be undone. So, it is very important that an adequate amount of shearing and heating take place in the extruder. However, too much shearing can decrease the strength and water holding ability of the final TVP product. Many years of research have determined the proper levels of shearing within the extruder barrel for meat extenders and meat analogs. The proper amount of moisture within this plasticized dough is also very important in increasing the potential for reaction sites between these proteins.

"Dies for TVP should have smooth streamlined flows that do not disrupt or cause shearing effects to the already laminated and cross-linked protein molecules" (Hagaiv et al., 1986). Some extruder manufacturers have developed TVP dies with Teflon® inserts to aid this streamlined uninterrupted flow. The internal appearance of some TVP dies appears as if they might have come from the aviation industry.

After the product leaves the extruder, it can be wet milled for proper sizing and/or dried.

TVP ADDITIVES

The addition of minor ingredients or the chemical adjustment of the TVP raw material can enhance various aspects of the finished product and lessen the specification constraints of some of the raw material.

The addition of calcium chloride can increase textural integrity and smooth the product surface of the TVP product. Dosage levels of calcium chloride range between 0.5 and 2.0%.

When supplementing light-colored meats with meat extenders made from textured vegetable proteins, it is desirable to bleach or lighten the color of the meat extender. Bleaching agents such as hydrogen peroxide range from 0.25 to 0.5%. Pigments such as titanium dioxide are also used at levels between 0.5 to 0.75% to lighten color; increased levels will weaken the textural properties of TVP. Color can be injected into the final sections of the extruder barrel (just before the die) to achieve special effects such as marbled, striped, and multicolored products. Colors can be sprayed onto the extrudate after it leaves the extruder to simulate striping found in bacon.

Increasing the pH of vegetable proteins before or during the extrusion process aids in texturization of the protein. "Extreme increases in pH increases the solubility and decreases the textural integrity of the final product" (Boison et al., 19829). "Modifying above pH 8.0 may also result in the production of harmful lysinoalanines" (Simonsky and Stanley, 1982). Lowering the pH has the opposite effect and decreases protein solubility making the protein more difficult to process. Sour, undesirable flavors may occur at pH below 5.0.

Sodium chloride or salt does not appear to have any beneficial effects on the texture or performance of textured vegetable proteins. Research has shown that the addition of salt actually seems to weaken textural strength. Sodium alginate addition can increase chewiness, water holding capacity, and density of extruded protein products.

Soy lecithin added to formulations of vegetable proteins at levels up to 0.4% tended to assist smooth laminar flow in the extruder and die which permitted the production of increased density soy products. The ability to make dense vegetable protein products is related to the higher degree of cross-linking that occurs during the extrusion process.

Sulfur, which is known for its ability to aid in the cleavage of disulfide bonding, assists the unraveling of the long twisted protein molecules. This reaction with the protein molecules causes increased expansion and smooth product surfaces and adds stability to the extrusion process. These benefits, however, are not without some undesirable side effects including off flavors and aroma. Usage levels are 0.01 to 0.2%.

The demand for meat extenders and meat analogs will continue to rise. Meat extenders still make up the largest segment of the textured vegetable protein market; however, the use of meat analogs is increasing. We are becoming more aware nutritionally of the foods we eat. Along with the beneficial high protein content of actual meats, there are some negative health benefits, namely, cholesterol. However, most people still like their meat. Meat analogs have become a viable alternative in offering a nutritionally acceptable meat substitute that in some cases comes close to matching actual meat products. Food scientists have made major headway in improving flavor, texture, mouthfeel, appearance, and color of meat analog products. In the marketplace, you can see more meat analog and meat extender products including bacon bits, soy burgers, and meat-free foods such as hot dogs, chicken nuggets, breakfast sausage patties/links, and bacon to name a few. Many of these products are even packaged in the same fashion as their meat counterparts.

READY-TO-EAT BREAKFAST CEREAL PRODUCTION

TYPES OF EXTRUSION-PROCESSED READY-TO-EAT BREAKFAST CEREALS

Extrusion cooking of ready-to-eat (RTE) breakfast cereals offers numerous processing advantages over conventional processing methods (Figure 6). Faster processing times, lower processing costs, less square footage required, shorter response time, and perhaps most importantly, greater flexibility are among the advantages. The added flexibility is

Figure 6 Examples of various extruded direct expanded and flaked breakfast cereals (from top row left to bottom row right): crisp rice, multicolored sugar-coated fruit rings, cornflakes, cinnamon and sugar graham shapes, cocoa rings, sugar-frosted strawberry stars, multicolored sugar-coated rainbow balls, and bran/wheat flakes.

made possible by the ability to quickly and easily change the extruder and/or processing steps for countless variations of RTE cereals. Simply changing the die, processing conditions, formulations, coatings, and coating additives will produce a wealth of different shapes, textures, and appearances of direct expanded RTE cereals, while the same holds true for flaked cereals.

Any and all types of predominately grain-based raw materials can be continuously processed on an extruder for RTE breakfast cereals. The extruder can be used as only a cooking step in the entire process and in others it can cook, shape, and process the final cereal product. After drying, the cereal can be packaged as is or blended with additives such as dried fruit, nuts, marshmallow pieces, etc., or sugarcoated and then packed.

Extrusion cookers are used extensively in the RTE cereal industry to cook doughs that are cold formed into various shapes and then gun or tower puffed. These products are what is called indirect expanded ready-to-eat breakfast cereals. In some instances, these types of cereals are referred to as half products because the cereal is only halfway through the extrusion process. The majority of expanded RTEs are still produced this way. This process starts with a flour blend that is mixed with other minor ingredients like sugar, salt, and color and then is metered into the extruder where it is preconditioned and cooked. A cooked dough ball is formed by a die and cut with a rotating knife. This hot dough ball is cooled and then fed into a single-screw cold former. The former cools and densifies the warm porous dough and forms it into a dense unpuffed half product or pellet that is dried to 9–11% moisture and then tempered for at least 24 hours prior to gun or tower puffing.

Gun or tower puffers heat the pellet to as high a temperature as possible under atmospheric conditions without scorching or burning. This heated pellet is then subjected to more heat under high pressure (typically using super-heated steam) and pulled into an expansion tank which is under a vacuum. These extreme differences in quick sequence of heat, heat/pressure, and vacuum cause the pellet to expand in a similar fashion as to how it would expand off the extruder. Further drying is usually not required after the puffing tower. In a puffing gun or tower, expansion of the pellet occurs in a three-axial plane. Extruders expand products in a double-axial plane. These differences in expansion created a few problems that took some time to overcome with die and extruder configuration technology when making direct expanded cereals and comparing them to gun or tower puffed cereals. Today, breakfast cereal producers are converting some tower puffed products to extrusion puffed and are introducing more new products that are extrusion direct expanded. The general trend in the RTE cereal industry is to eliminate puffing guns and towers and to slowly move to all direct expanded products.

As can be seen by Figure 7, the direct expanded cereal process is a lot less complicated than the indirect expanded method and takes less processing time and manpower. Direct expanded breakfast cereals are those products which are cooked, directly expanded and shaped, and cut off the extruder and then dried and/or coated and packaged.

All extruded direct expanded breakfast cereals can closely match their nondirect expanded counterparts except for oat-rings cereals. Extruded flaked cereals also can closely match traditionally processed flaked products. Ingredient and nutritional labeling on the box would be the same regardless of how the cereal was processed, again, except for oat rings. The difficulties encountered with DX extruded oat rings is discussed later.

The raw materials for traditionally produced flakes are usually cooked in large pressured rotary cookers. After these cook, the cooked material is graded and slowly dried and then tempered, flaked, dried/toasted, and packaged. These processes use huge factories and re-

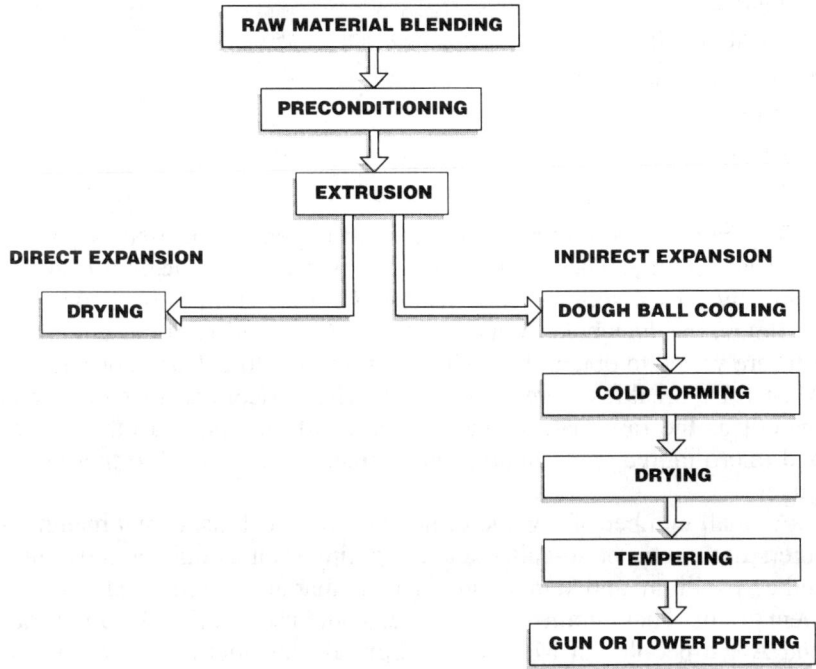

Figure 7 Comparison of extrusion-cooked indirect and direct expanded ready-to-eat cereal products.

quire, in some cases, 24-hour processing times. You can see why some producers are switching over to extruded RTE breakfast cereals.

The most commonly used cereal grains in the ready-to-eat breakfast cereal industry are rice, wheat, corn, and oats. Granulation of these components can be varied.

COMMONLY USED RAW MATERIALS FOR RTE BREAKFAST CEREALS

The following are commonly used raw materials in the production of RTE breakfast cereals:

- yellow corn flour, meal, and cones
- patent of clear wheat flour
- whole wheat flour
- heavy and refined wheat bran
- white and brown rice flour and meals
- whole and defatted oat flour
- sugar
- wheat starch
- sodium bicarbonate
- colors
- barley malt extract
- salt

Typically a blend from the above listed ingredients is used as the raw material mixture. These ingredients can be from organically or nonorganically grown sources. Raw materials milled from these grains can be flours, meals, whole flours, and brans. Granulation of these ingredients are varied to obtain the desired texture, mouthfeel, and appearance. Most RTE breakfast cereals use corn, wheat, rice, and oats or blends thereof as the raw material base. Sugar, salt, colors, and other minor and micro ingredients (vitamins and minerals) are added to these base mixes.

A small number of organic or health food breakfast cereal manufacturers use exotic or so-called ancient grains such as quinoa, amaranth, millet, psyllium, and kamut. Quinoa and amaranth are referred to as ancient grains. These grains were cultivated and used by the Mayan, Incan, and Aztec peoples of Mexico and Central and South America. For a grain, amaranth is relatively high in protein and Vitamin E. Millet is a small seeded cereal grass. Psyllium is grown in India and used in the RTE breakfast cereal industry as a high fiber source. Kamut is a large kernel wheat that is cooked whole and flaked to produce some types of wheat/bran flakes.

Certain formulations can be protein fortified using various protein sources such as soybean, sesame, dried milk, and whey. Some of these protein-rich constituents must be used sparingly due to the negative flavor and texture impact they may have.

Regardless of the ingredients used, the process basically stays the same when making direct expanded or flaked extruded RTE breakfast cereals. As one can see, a quick and simple way to make many product changes is to vary the recipes.

Yellow corn cones are milled from yellow hard vitreous horny endosperm corn that has had the germ and bran removed and ground and sized and sifted to the proper granulation. Breakfast cereal manufacturers throughout the world are very specific on the varieties of corn milled for their use. Corn cones are classified as fine meal, its granulation falls between flour and meal. Cornmeal is the next coarser grade from corn cones. Cornmeal is also referred to as snack meal, 36 grits or brewers grit or meal. There are also grades of coarse meal and flaking grits. Flaking grits are half pieces of the corn kernel with the bran and germ removed that are used in producing traditional cornflakes.

Typical Analysis of Yellow Corn Cones.[a]

Moisture	11.0–13.0%
Protein	6.0–8.0%
Ash	0.2–0.3%
Fat	0.6–1.0%
Fiber	0.25–0.4%
Starch	76.0–81.0%

[a]Courtesy of Illinois Cereal Mills, Inc., IL.

Yellow corn cones are used in conjunction with a barley malt extract slurry when producing extruded corn flakes. The barley malt extract is a liquid blend of sugar, water, barley malt extract, and salt. This slurry is injected into the extruder barrel. The corn cones are preconditioned separately prior to extrusion and then brought together with the barley malt slurry in the extruder barrel. If the dry sugar and salt were mixed with the corn cones, they would compete with the moisture in the preconditioner. Sugar and salt hydrate much faster than the cones, so the corn cones would be poorly prehydrated in the preconditioner.

To vary the texture, appearance, and mouthfeel of extruded cornflakes, a blend of yellow corn flour, cones, or meal can be used. When using corn flour, the resulting extruded flake becomes less blistered with a smooth, flat surface. As you introduce cornmeal into the formula, the flakes become more blistered. Using too much corn flour will result in completely flat and smooth, unappealing flakes. Using too much cornmeal causes the flakes to blow up into what some people in the industry call "fish eggs" or "Easter eggs." This over-puffing is caused in the oven when the large starchy particle in the wet flake expands. The inverse is true when using corn flour. Using around 5 to 10% of either cornmeal or flour with corn cones can produce an acceptable cornflake.

Wheat flour used in the RTE breakfast cereal industry is milled from soft or hard white wheat. Red wheat varieties are not used. Patent flour is the higher grade of wheat flour, as much of the germ and bran have been removed as is physically possible with modern milling practices. This grade has the lowest ash and fiber content. Clear flour is a lesser grade of flour and typically contains more fiber and has a higher ash value. Patent flour is creamy white in color, and clear flour has more of a light gray hue. Patent flour is the grade of wheat flour normally used in the breakfast cereal industry.

Typical Analysis of Patent Wheat Flour.[a]

Moisture	14% maximum
Protein	8.5–10%
Ash	0.43–0.47%

[a]Courtesy of King Milling Company, MI.

Patent soft or hard white wheat flour is blended with yellow corn flour, oat flour, sugar, salt, and color, and this mixture is used to produce the extruded direct expanded rings and balls that are so popular in the cereal industry. These direct expanded products are then dried and sugar- and flavor-coated.

When whole wheat flour is required in the formula, the entire kernel of hard or soft white wheat is ground to flour.

Typical Analysis of Whole Wheat Flour.[a]

Moisture	14% maximum
Protein	9.5–10%
Ash	1.6% ± 0.2%

[a]Courtesy of King Milling Company, MI.

Whole wheat flour is commonly used in products like bran flakes, wheat flakes, bran sticks, and bran nuggets. Whole wheat flour is blended with heavy bran and/or refined bran, sugar barley malt extract, and salt for these products. Heavy bran is produced on a special milling process that "butterflies" the wheat kernel and leaves approximately

70% of the endosperm with the bran layer. Because heavy bran is a coarse product, the resulting product produced with it has a very appealing, rough, natural look. Refined bran is a cleaned bran in which as much of the endosperm and germ have been removed as possible. Refined bran is used in high fiber formulations. A combination of brans and flours can produce the product desired.

Typical Analysis of Heavy Bran.[a]

Moisture	14% maximum
Crude fiber	4% average
Total dietary fiber	13% average

[a]Courtesy of Knappen Milling Company

Typical Analysis of Refined Bran.[a]

Moisture	14.5% maximum
Crude fiber	11% maximum
Total dietary fiber	39.4% average
Protein	13–15%

[a]Courtesy of King Milling Company, MI.

White rice flour is used to produce crisp rice, rice flakes, and other products. Brown rice flour and/or meal can also be used for these products. White rice flour and meal have the bran and germ removed in the milling process. Like cornflakes, crisp rice and rice flakes use a separate barley malt slurry that is pumped into the inlet head of the extruder barrel. Because rice flours and meals are good expansion agents, they can be added to formulas that might need a good expanding agent. Producing direct-expanded extruded crisp rice is a simpler, less complex processing method than that used by Kellogg to make Rice Krispies®. Rice Krispies® are made by pressure cooking whole pearled kernels of white rice with a barley malt slurry. The cooked rice is then dried, bumped, tempered, and puffed in a high-temperature oven.

Typical Analysis of White Rice Flour.[a]

Moisture	10–12%
Protein	6.5–8.5%
Fat	0.5–1.2%
Ash	0.4–0.8%
Crude fiber	0.3–0.7%

[a]Courtesy of Riviana Foods, TX.

Because whole oat flour is higher in protein and fat than most all cereal grains, it presents special problems when extruding. It is the most difficult of all the cereal grains to extrude. This high protein and especially the fat affect expansion. Extruding oat rings, which have whole oat flour of 50 to 85% in the formula, can be a trying experience. The addition of a good neutral flavored expanding agent like raw wheat starch is used to counter the poor expansion effects of whole oat flour. In turn, though, this dilutes the pleasant oat flavor associated with the whole oat flour. If a better expanding agent was not added to the flour due to the high fat and protein, the final product would be hard and would have a straight flat side, not the donut-like shape desired. Because of this, the leading oat rings cereal, Cheerios®, made by General Mills Corporation, is produced on the indirect expansion process.

There are commercially available direct expanded oat rings, but generally, they are the cheaper, so-called store brand products.

Recently oat millers developed a process to extract most of the fat from whole oat flour. When using defatted oat flour, the same adverse processing problems allied with full fat whole oat flour are not encountered. As with most foods, though, the fat usually carries the flavors, so, oat flavor might be lacking when using defatted oat flour.

Sugar or other sweeteners like honey, malt, molasses, and dextrose are obviously used as sweeteners. They also control expansion, harden the cereal's texture, and improve bowl-life. Salt also helps control or limit expansion and enhance flavor. Sodium bicarbonate is added as a leavening agent. It also assists in the browning (Maillard reaction) process and enhances flavor.

Ready-to-eat (RTE) Breakfast Cereal Extrusion Operation

There are not as many complex modifications taking place during the extrusion of RTE breakfast cereals as during TVP production. The primary modification happening in the extruder is gelatinization of the starchy components of the raw material. Most cereals contain substantial amounts of starches that are cooked to develop desired textures, colors, flavors, and other physical properties. Cooking or gelatinization of starch in the traditional cereal process is controlled by time, temperature, and availability or presence of water. In the extrusion process, shear is a fourth dimension that impacts product quality.

The extruder screw profile or configuration is specifically designed to produce RTE breakfast cereals. The use of a preconditioner prior to use of the extruder is not always necessary. Typically, direct expanded cereals without bran in the formula do not require preconditioning, while direct expanded products with bran and all flaked cereals usually need preconditioning to make a satisfactory product.

As mentioned earlier, when producing most flaked products, there is a separate barley malt slurry that is injected into the extruder barrel. The dry raw materials are preconditioned separately. Many producers have found that doing this ensures that the dry-grain-based raw materials are adequately premoistened in the preconditioner before the extruder barrel. Because the salt and sugar are so hygroscopic, they are handled in a slurry form and do not "rob" moisture from the starch (Figure 8).

Extruding direct expanded RTE cereals is fairly straightforward: precondition (in some cases as mentioned above) extrude, determine size and shape at the die, dry and package or sugar coat and package. Some DX cereals are produced on single-screw extruders but most are made on twin-screw extruders (Figure 9).

The task of selecting a proper extrusion process to yield a desirable flaked breakfast cereal involves consideration of three key areas: raw material selection, hardware components or system configuration, and software or processing conditions. There are numerous "schools" of thought in how extruded, flaked RTE breakfast cereals can be manufactured. Outlined below are four different methods for making an extruded, flaked breakfast cereal (Figure 10).

It has been proven that extrusion configuration one makes the best flaked breakfast cereal. There is currently considerable research going on with testing configuration four. This research is showing favorable results with all flaked and direct expanded cereals. Configurations one, two, and three will produce flaked and expanded RTE breakfast cereals. It has been determined that preconditioning is necessary to produce an acceptable cornflake. Keeping this in mind, configuration two would be a poor choice. Cornflakes are still considered "king" of the RTE breakfast cereal market.

What exactly is a "good" cornflake? This is a tough question to answer as many factors enter in. Bowl-life is always mentioned when critiquing any breakfast cereal. It has been recognized that without preconditioning, an extruded cornflake has a poor bowl-life. In other words, it gets soggy too fast when it is in milk. When analyzing any breakfast cereal, shape, appearance, texture, density, color, molar impaction, and bowl-life are mentioned.

Depending upon the method employed, a flaking bead is either cut at the extruder, former, or sizing device. After this step, the flaking beads need to be conditioned. Bead conditioning systems cool the outside of the bead, temper the bead, and assure that only single beads are passed along to the flaking rolls. The flaking rolls flake the bead into the characteristic flaked shape and then the wet flake is dried and toasted. Sugar coating can be an option after drying/toasting.

Sensory evaluation techniques and physical structural analyses have

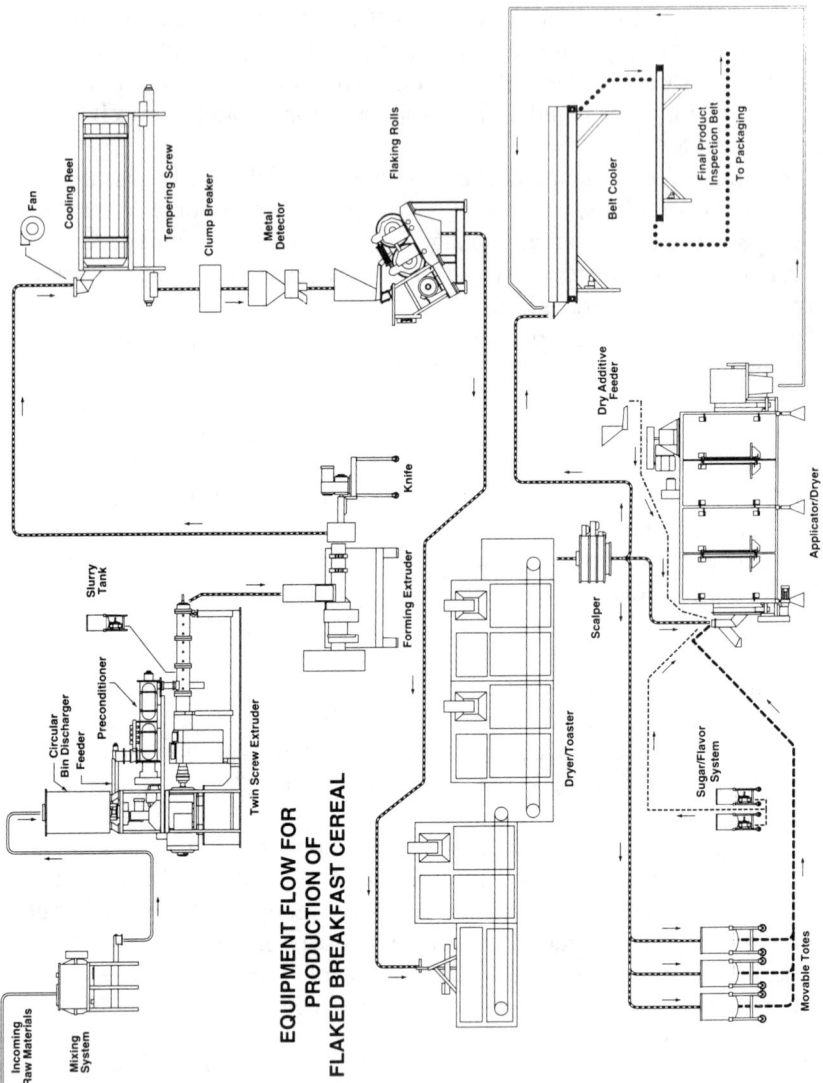

Figure 8 Equipment flow for production of flaked breakfast cereals.

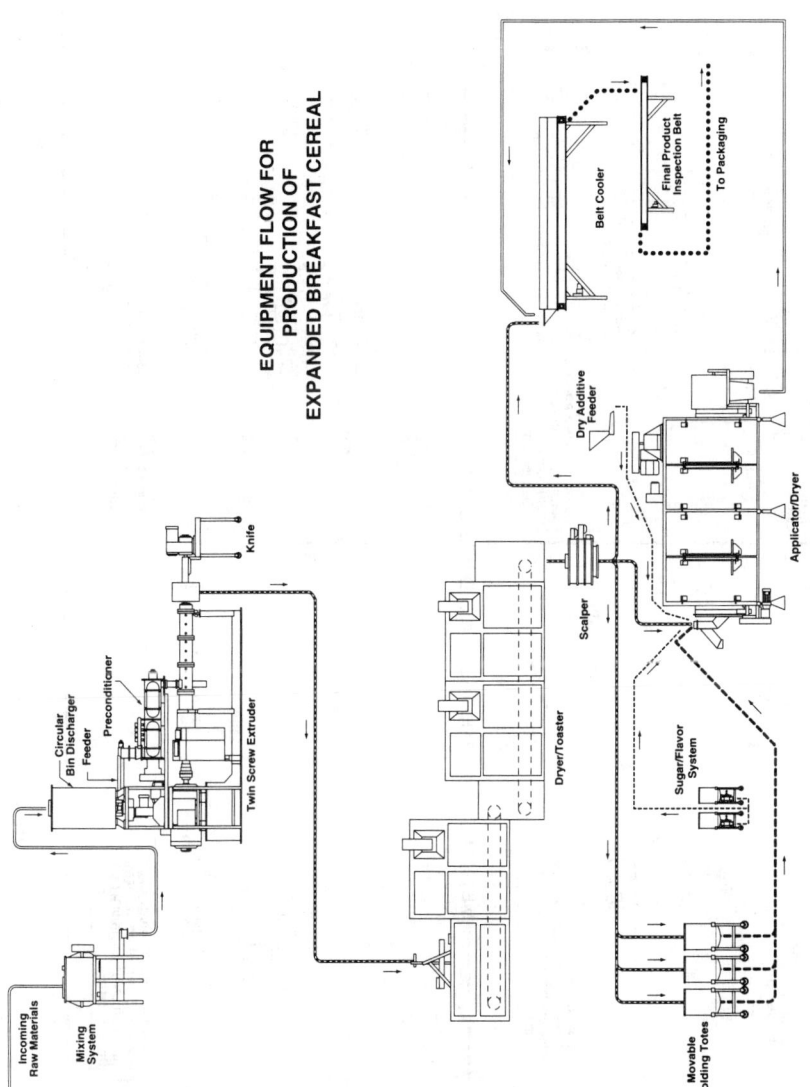

Figure 9 Direct expanded RTE breakfast cereal flow.

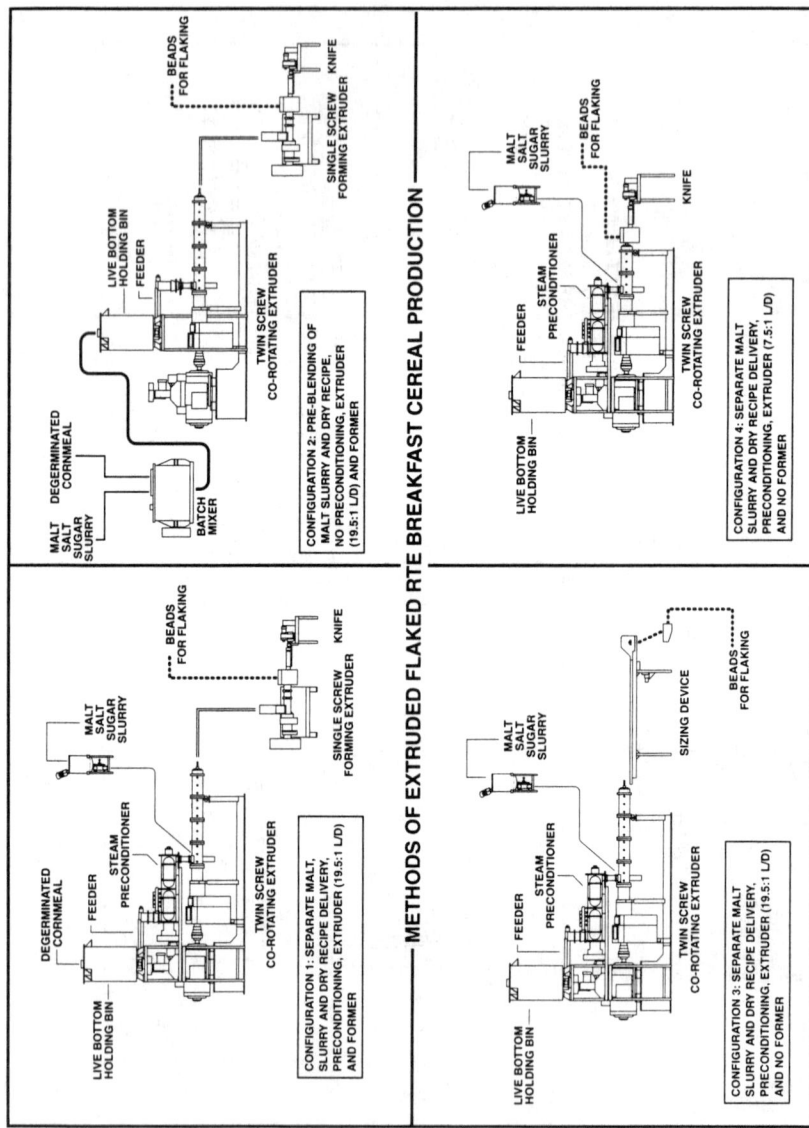

Figure 10 Methods of extruded flaked RTE breakfast cereal production.

indicated differences in flakes and direct expanded cereals depending on the extrusion configuration used. However, it is possible to extrude flakes and direct expanded cereals with flavor and textural attributes similar to those in the traditional process through proper selection of extruder configurations and management of processing conditions.

Extrusion offers a less costly viable alternative to traditionally processed cereals. The future of RTE cereals points in the direction of extrusion.

DIRECT EXPANDED (DX) AND THIRD GENERATION (3G) SNACKS

Consumers today are demanding ever-broadening selections of a variety of snack foods. Extrusion has provided a means of manufacturing new and novel products and has revolutionized many conventional snack manufacturing processes. Extrusion equipment offers many basic design advantages that result in minimizing time, energy, and cost while at the same time increasing the degree of versatility and flexibility that was not previously available.

Many common snack foods such as fired or baked collets (corn curls) are corn based and are produced on rather simple single-screw, high pressure extrusion cookers (Figure 11).

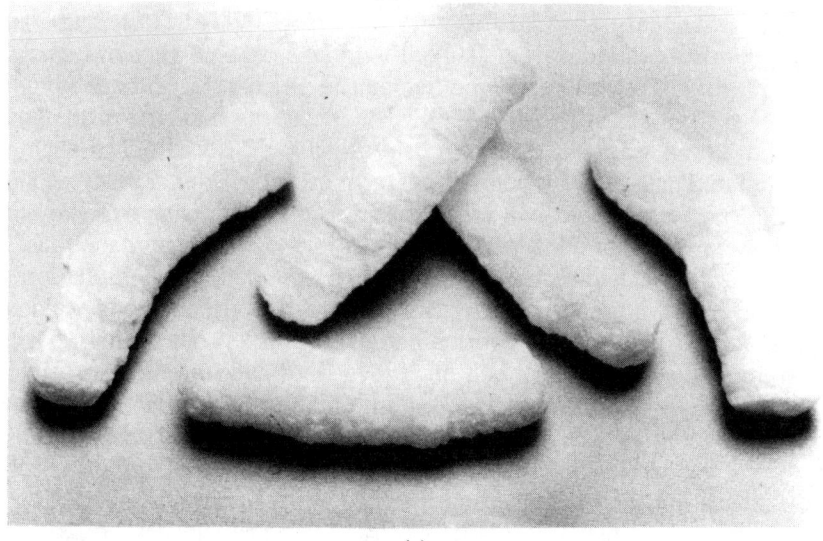

(a)

Figure 11 Various direct expanded snacks—corn curls, balls, and rings.

(b)

(c)

Figure 11 Various direct expanded snacks—corn curls, balls, and rings.

Other snack products (for example, half products or third generation products) may require more sophisticated twin-screw extrusion equipment (Figure 12).

DIRECT EXPANDED SNACKS

The majority of extruded snacks on the market fall into the category of direct expanded snacks. And, the most popular one of these is cornmeal based.

Properly selected cornmeal is fed into an extruder with a feeding device that delivers the meal at a constant rate. The meal is exposed to moisture, heat, and pressure as it is transported through the extruder into the extruder die. It has been said that direct expanded corn snacks are the easiest food products to run on an extruder.

Extruders for direct expanded snack products are normally short in length (less than 10:1 L/D). It is important that the extruder configuration consisting of screws, steamlocks, and barrel segments be properly selected to feed, knead, and cook the process material as it passes through the extruder.

Moisture of the direct expanded snack is normally between 8 and 10% moisture content on a wet basis and requires additional drying to

Figure 12 Various 3G snacks, puffed and unpuffed.

produce the desired product crispness. With drying, this moisture is brought down to 1 or 2%.

The most popular flavoring for direct expanded extruded snacks is cheddar.

The specifications of cornmeals used tend to influence the texture and mouthfeel of the final snack product. Many grades or granulations of degermed cornmeal are available.

Typical Analysis of a Cornmeal for Softer Textured Snack.

Moisture	11.0%
Protein	8.0%
Fat	0.8%
Fiber	0.4%
Ash	0.3%
% granulation on U.S. sieve	
On 30	0.1%
On 40	0.4%
On 60	80.0%
On 80	15.0%
Thur 80	4.5%

Typical Analysis of a Cornmeal for a Crunchier Textured Snack.

Moisture	11.0%
Protein	8.0%
Fat	0.8%
Fiber	0.4%
Ash	0.3%
% granulation on U.S. sieve	
On 10	0%
On 14	2.5%
On 20	65.0%
On 40	31.0%
Thur 40	0.5%

For example, if a finer texture cell structure or softer bite for a snack is desired, a snack producer may want to use materials having an analysis as shown in the first cornmeal chart above. Or, if a crunchier product is desired, meal with the specifications found in the second chart would be used (Figure 13).

THIRD GENERATION SNACKS

Third generation (3G) snack products or pellets are sometimes referred to as semi or half products. Following extrusion cooking and forming, these hard dense pellets are dried to a stable moisture content

Direct Expanded (DX) and Third Generation (3G) Snacks

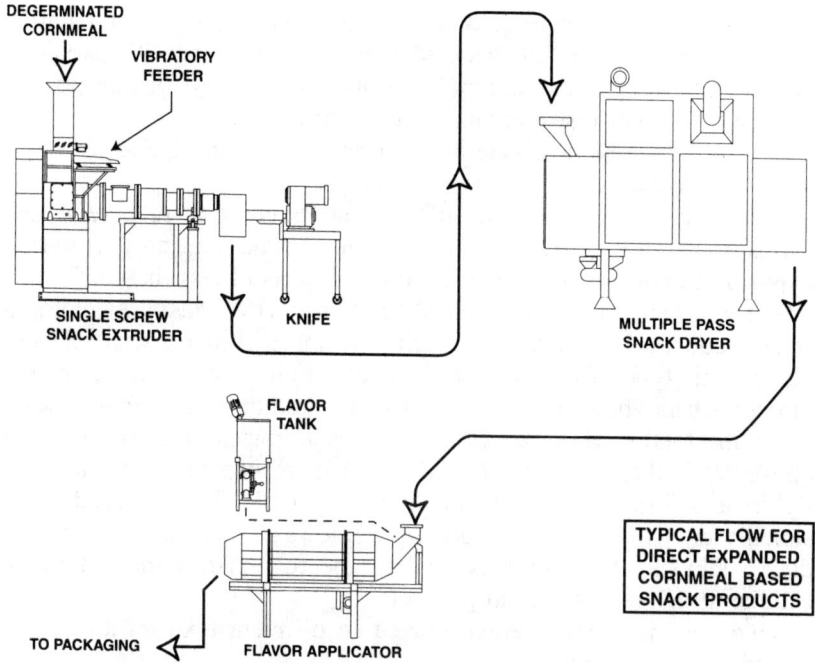

Figure 13 Typical flow for direct expanded cornmeal-based snack products.

assuring shelf stability. When dried, the pellets may be distributed to a snack processor where they are expanded by hot oil, hot air puffing, or microwaving (Figure 14).

The expanded products are then seasoned with salt and various spices, packaged, and sold to consumers as ready-to-eat (RTE) snacks.

Figure 14 Third generation snack pellet, showing phases of expansion in hot air.

3G pellets may also be sold directly to the consumer for preparation in the home. This type of snack adds new dimensions to its marketing potential because of its high bulk density and stability for transporting to various smaller distribution sights (Figure 15).

A wide range of raw materials can be selected and blended to produce excellent recipes for many 3G snacks.

Generally, the combination of ingredients contains relatively high levels of starch to maximize the expansion of the final product during exposure to hot oil, air, or microwaving. Levels of less than 60% total starch in a recipe result in reduced final product expansion, yielding a final product with increased crunchiness and a firmer texture. Examples of this type of product are those containing mainly whole cereal grains such as whole ground corn or masa in traditional corn chip snacks.

As the total starch level in a third generation product is increased above 60%, the final product will yield more expansion resulting in lighter and softer textures. Shortenings, vegetable oils, salts, and occasionally emulsifiers are included in recipes as processing aids, to reduce stickiness, to control expansion, and to impart a more uniform cellular structure in the final product.

Two generic recipes representing two different textures are represented by the following:

Typical Formulations for Third Generation Snacks.

Hard Crunchy Texture		Soft, Light Texture	
94.5%	Ground corn	56.0%	Corn starch
5.0%	Corn starch	27.5%	Wheat starch
0.5%	Monoglyceride	14.0%	Tapioca starch
		2.5%	Liquid vegetable shortening

Monoglycerides may significantly reduce expansion during secondary puffing if used at levels of 0.5% or above. Eastman Chemical's Myvaplex 600 or Grinstead's Dimodan P.V. are examples of emulsifiers.

Careful consideration must be given in the selection of cereal grains, starches, proteins, and other minor ingredients when designing recipes for such products. For example, starch has many contributions to the final product including expansion, flavor, binding, viscosity development, caloric value, resilience, and functionality (for example, hardness, texture, etc.).

It is also important to consider ingredients available in your local area because ingredient cost is the major expense of producing such products. For example, in the U.S., corn or wheat are major starch sources, while in Europe, potato is a major starch source, and in Asia,

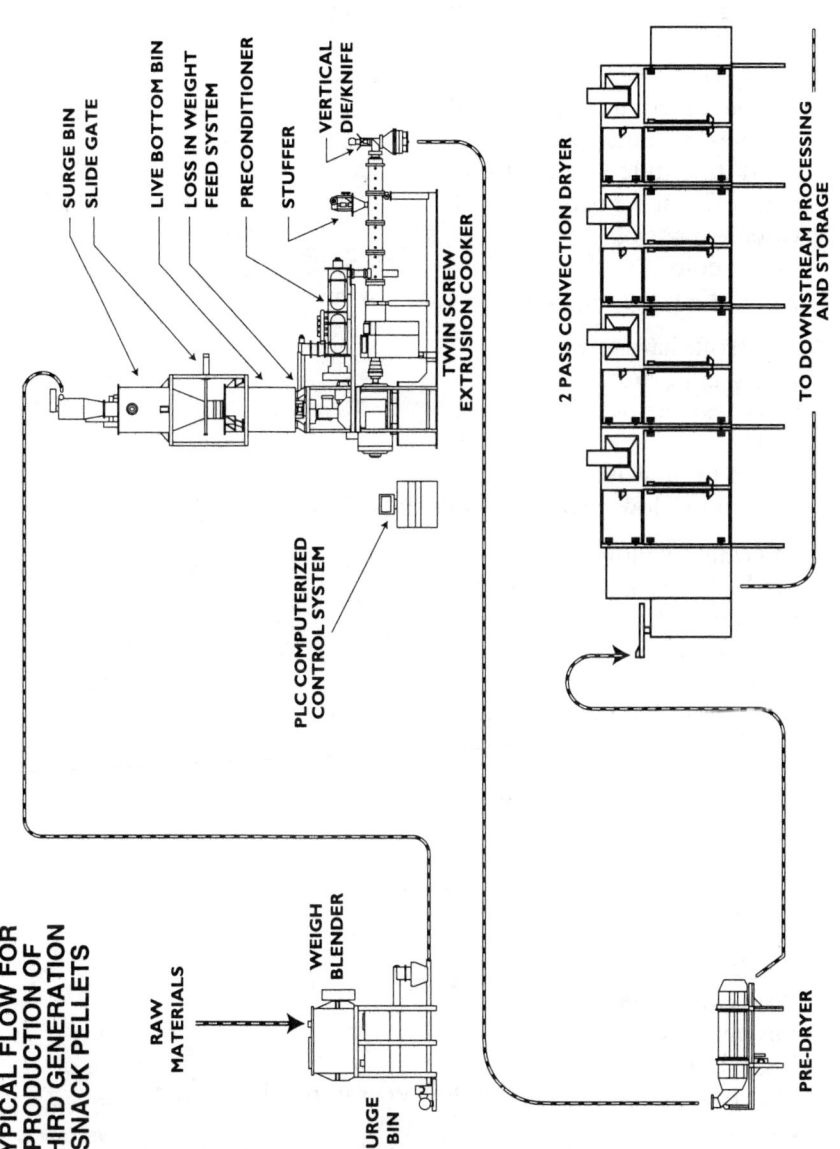

Figure 15 Typical flow for production of third generation snack pellets.

tapioca or rice is a major starch source. It is also important to consider each individual starch source and their strong points.

Cereal grain starches
   Rice (long, medium, short grain)
      Smallest granule size
      Most digestible when cooked
      Bland flavor—easily flavored
      Good expansion
      White color
      Highest energy requirements for cooking

   Wheat (soft, hard, durum)
      Fairly large granules
      Good expansion
      Mild flavor
      White to off-white color
      Medium to low energy requirements for cooking

   Corn (white, yellow)
      Medium size granules
      Good expansion
      Definite flavor
      Yellow color
      Medium to high energy requirements for cooking

   Barley
      Medium to large size granules
      Fair expansion
      Light brown to gold color
      Low energy requirements for cooking

   Oats
      Small size granules
      Poor expansion
      Strong flavor
      Light brown color
      Starch portion has relatively low energy requirements for cooking
         Because of high lipid content, higher amounts of energy input are required for cooking.

Tuber starches (lower in protein)
   Potato
      Very large starch granules
      Develops very high viscosity when cooked

Excellent swelling power
Excellent binder
Definite flavor
Gold to light brown color
Low energy requirements for cooking
Granules breakdown easily

Tapioca
Medium size starch granules
Develops high viscosity
Excellent binder
Bland flavor
White color
Low to moderate energy requirements for cooking

Many types of proteins and protein enrichments may be added to third generation snack-type recipes such as meats (whole fresh shrimp, fresh chicken, beef, etc.), dairy products (cheese, yogurt, milk solids), and legume proteins (soy, pea, bean). Up to 30–35% levels may be added and still maintain high quality final products.

Several minor ingredients have very useful effects on the texture, quality, and flavor of the final products. Salt is very useful in assisting with uniform moisture migration throughout the third generation pellet after drying during the moisture equilibration period. Baking soda will give special flavor and textural attributes to the finished products after frying, puffing, or microwaving. Oils or emulsifiers reduce stickiness during cutting and other processing steps.

After the dry ingredients have been uniformly blended, it is suggested that the liquid ingredients such as shortenings, flavors, or water be applied as a spray in the batch mixer or be injected into the preconditioner of the extruder in the cooking extrusion system. During the extrusion cooking step, it is vital that the raw materials be completely cooked (unless the recipe contains pregelled starches) if the objective is to maximize the final product expansion.

A good cook is defined as the combination of temperature, residence time, and moisture content during extrusion to fully gelatinize the starchy components in the recipe. The temperature profile in the extruder (which is dependent upon the ingredient characteristics, extruder configuration, and processing conditions) will be higher than the gelatinization temperature of the starches used in the recipe unless pregelled starches are used.

Typical extrusion cooking processing conditions may vary depending on the type and amount of starches used in the recipe. Generally, temperatures in the cooking zone of the extruder will range between 80 to 150°C (175 to 302°F), and barrel temperatures in the forming zone will range between 65 to 90°C (150 to 194° F). Extrusion moistures will range between 25 to 30% moisture content with a residence time of 30 to 90 seconds.

The total energy requirements for producing third generation snacks may be quite low when using tuber starches or pregelled starches in your recipe. However, when using whole cereal grains or high protein wheat flours, increased mechanical energy is desirable to fully gelatinize the recipe and to lower the molecular weight of the starch granules to allow improved expansion characteristics during frying, hot air puffing, or microwaving.

Thermal energy sources include the use of steam, hot water, or other thermal fluids circulated through the jacketed barrel to provide heat transfer into the extrudate (external heating) and/or steam and hot water injected directly into the product in the preconditioner or extruder barrel. The cooking extruder has segmented barrels (heads) and screws for product versatility. Various head, screw, and steamlock designs can be incorporated into the configuration to produce the desired cooking conditions.

For example, when using recipes consisting mostly of tuber starches, only low shear screw configurations are required to make pellets that will yield a light-density, soft-textured final product. However, if large amounts of wheat flour are used, screw configurations designed to impart higher levels of mechanical energy into the pellets are required to make the same light-density, soft-textured final product.

Following this cooking step, the material is passed into a venting zone and then into a forming extrusion zone that cools and densifies the cooked, plasticized mass. This forming step may be accomplished in a separate extruder or in a secondary zone of the cooking extruder.

This forming portion of the process contains a low shear screw configuration containing a minimum of restrictions except for the final die. The forming zone is usually characterized by a positive, forward transport screw configuration complemented by maximum cooling to reduce product temperature to 70 to 95°C (158 to 203°F).

The cooled, viscoelastic product is then shaped by the final forming die that contains sufficient open area to prevent excessive pressure buildup and, thus, expansion of the cooked dough. Many different shapes and sizes of third generation snacks may be produced. The cooked, densified, shaped product extruded through the former die contains between 20 and 28% moisture after a slight "flash-off" during the small decompression step.

For multidimensional snack products, the viscoelastic dough is formed into sheets that are sized to the proper thickness and then embossed with a desired pattern to give a specified design to the surface of the product. The sheets are then brought together and passed through a rotary die cutter that cuts the product into the desired shape. In many cases, the multicolored products are made with two extruders feeding one common die. This makes it possible to laminate two sheets with different recipes; therefore, two completely different textures may be incorporated into the same product piece. Stenciling or imprinting logos, designs, or wording on the sheets may be done for uniqueness.

Proper drying is required to reduce the moisture content of the pellet to approximately 6–8% moisture which is most suitable for frying. The drying step is a very critical step in the production of good quality third generation snacks. Drying temperatures of 70 to 95°C (158 to 203°F), humidity control, and retention times of one to three hours are employed to accomplish this important processing step. Immediately after exiting the dryer, some moisture in the extruded pellet is distributed close to the periphery of the product, but most of it is still located in the central points of the pellet. Various studies of this procedure usually involve complete immersion of the pellet in 150 to 200°C temperatures, and experience shows that properly dried third generation snack pellets fry up or expand more uniformly following an equilibration period of one or two days during which time the moisture migrates to reach an equilibrium within the final product.

Prior to consumption, third generation snacks are fried in hot oil, expanded in hot air, hot salt, a microwave, or an infrared oven. The frying procedure usually involves complete immersion of the pellet in 150 to 200°C (302 to 390°F) oil for 10 to 40 seconds (depending on the product recipe). The expansion of the pellets with hot air, infrared heating, or sand or salt puffing has gained popularity due to the low caloric content in the final product.

The frying process can be divided into three phases. During the first phase, there is moisture loss from the periphery of the snack pellet into the oil. The snack pellet becomes warmer and more plastic in texture. Heat penetrates to the center of the pellet, and the moisture distributed throughout the pellet turns to steam. If the product was thoroughly cooked and properly formed, it will have gas retaining properties that are critical to the textural development of the expanded product. The moisture in the product will evaporate due to the high temperature of the oil and will expand inside the product where it is trapped. The result is production of an expanded product having fine cell structure with desirable mouthfeel and texture.

Extrusion systems for the production of third generation snacks and multidimensional third generation snacks are efficient, economical to run, and produce a product with built-in marketing flexibility due to long shelf life and high bulk density prior to frying or puffing.

REFERENCES

Boison, G., M. V. Taranto, and M. Cheryan. 1982. *J. Food Tech.* 18: 719.

Central Soya Chemurgy Division. 1990. *The Protein Book: A Guide to Soy Proteins from Central Soya*, p. 12.

Hagaiv, R. C., S. R. Dahl, V. J. Feron, and L. VanBeek. 1986. *J. Food Science.* 51: 367.

Rokey, G. J. 1990. "Wenger Mfg., Inc., Sabetha, KS, Process Description Textured Vegetable Protein" p. 215.

Simonsky, R. W. and D. W. Stanley. 1982. *Canadian Institute of Food Science and Technology Journal.* 15: 294.

Smith, O. B. 1975. "Products of textured soy proteins." Presented at the 1st Latin American Soy Protein Conference, Mexico City, Mexico, November.

Appendix

MASS AND ENERGY EVALUATION IN EXTRUSION SYSTEMS

I. Introduction to Mass and Energy Balances
   A. A mass balance will help you:
      1. Calculate the total mass flow at specific points in the system
      2. Calculate the moisture content, fat content, or other component content at a specific point in the system
      3. Calculate flow requirements of input streams to reach a specific component content
      4. Calculate mass flow requirements for a scaled-up system
   B. An energy balance will help you:
      1. Calculate the energy content of the product at a specific point in the system
      2. Calculate the temperature of the product at a specific point in the system
      3. Calculate energy requirements or consumptions of the process
      4. Often eliminate experiments because the process is not thermodynamically possible

C. In essence, a mass or energy balance is: In − out = accumulation
   1. In most cases, we are working with steady state processes, therefore, there is no accumulation in the system
   2. "System" may include:
      a) Preconditioner
      b) Extruder barrel
      c) Coating system
      d) Any combination of the above

II. Mass Balance Calculations
   A. Mass balance is:

   Using $\dot{m}$ to indicate mass flow per unit time $\left(\text{such as } \dfrac{kg}{hr}\right)$

   $$\sum \dot{m}_{in} - \sum \dot{m}_{out} = \Delta m_{sys}$$

   where: $\sum \dot{m}_{in}$ = Sum of all incoming mass flows
   $\sum \dot{m}_{out}$ = Sum of all outgoing mass flows
   Δm_{sys} = Accumulation of mass in the system

   1. Incoming mass flows:
      a) Steam injection ($\dot{m}_s$)
      b) Water injection ($\dot{m}_w$)
      c) Dry recipe ($\dot{m}_{dr}$)
      d) Fat (lipid) injection ($\dot{m}_f$)
      e) Other additives
   2. Outgoing mass flows include:
      a) Steam lost at preconditioner/extruder transition ($\dot{m}_{slp}$)
      b) Steam lost at vent ($\dot{m}_{slv}$)
      c) Product exiting die ($\dot{m}_{prod}$)
      d) Steam flashed at die ($\dot{m}_{sld}$)
   3. Accumulation of mass in the system:
      a) Normally considered to be zero because we want to operate at steady state
      b) Must allow time for equilibration before taking data
      c) Control loops on all mass flows are necessary to ensure steady state

   B. Application of Mass Balances—Mass Balance in the production of Salmon Feed (the composition of the material in the extruder barrel just behind the die must be determined. The example process with all of the known parameters is shown in the figure on the next page. This process will be used again for an energy balance example.)
      1. Calculate the moisture content

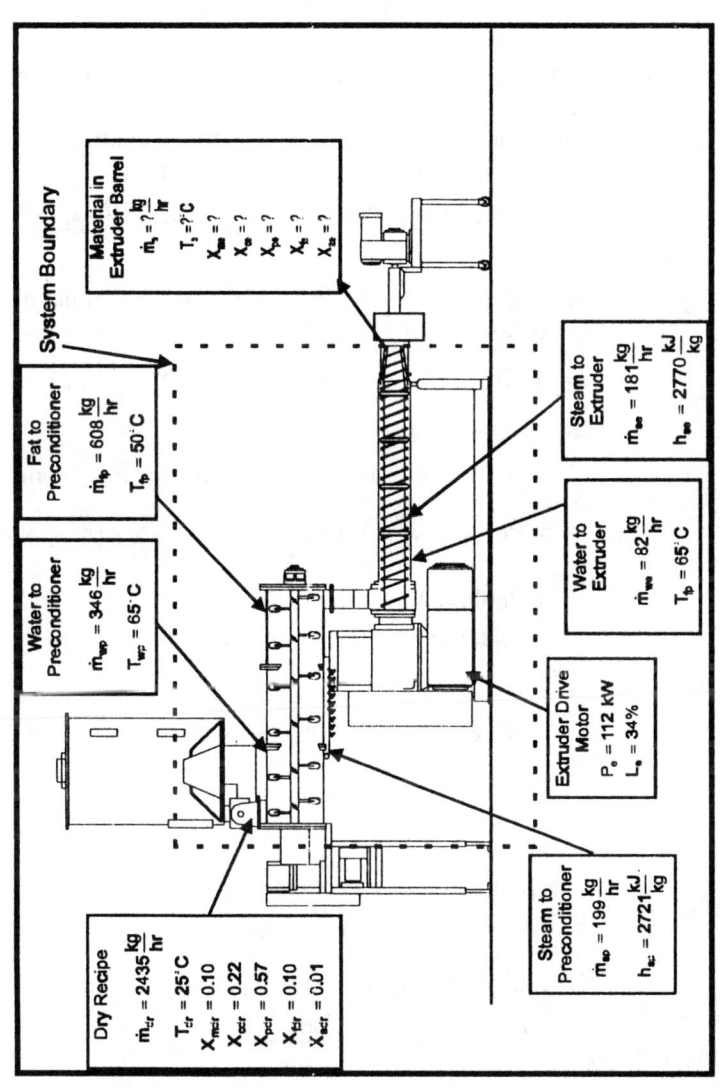

$$\sum \dot{m}_{in} - \sum \dot{m}_{out} = \Delta m_{sys}$$

$\dot{m}_{dr} \cdot X_{mdr} + \dot{m}_{wp} + \dot{m}_{sp} + \dot{m}_{we} + \dot{m}_{se} - \dot{m}_{slp} - \dot{m}_e \cdot X_{me} = 0$

Use X to indicate the mass fraction of a specific component.
In this case, X_m indicates mass fraction of moisture or water.
Values for X will always be less than or equal to 1.0.
Assume $\dot{m}_{slp} = 0$, solve the above equation for X_{me}:

$$X_{me} = \frac{\dot{m}_{dr} \cdot X_{mdr} + \dot{m}_{wp} + \dot{m}_{sp} + \dot{m}_{we} + \dot{m}_{se}}{\dot{m}_e}$$

Since we do not know the value of $\dot{m}_e$, we must do a total mass balance to determine it.

$$\dot{m}_{dr} + \dot{m}_{wp} + \dot{m}_{sp} + \dot{m}_{fp} + \dot{m}_{we} + \dot{m}_{se} - \dot{m}_{slp} - \dot{m}_e = 0$$

Again, assume $\dot{m}_{slp} = 0$, and solve for $\dot{m}_e$:

$$\dot{m}_e = \dot{m}_{dr} + \dot{m}_{wp} + \dot{m}_{sp} + \dot{m}_{fp} + \dot{m}_{we} + \dot{m}_{se}$$

$$\dot{m}_e = 2{,}435 + 346 + 199 + 608 + 82 + 181 = 3{,}851 \left(\frac{kg}{hr}\right)$$

Now, calculate X_{me}:

$$X_{me} = \frac{2{,}435 \cdot 0.10 + 346 + 199 + 82 + 181}{3{,}851} = 0.27$$

The moisture content in the extruder barrel is 27% wet basis.

2. Calculate the carbohydrate content

$$\sum \dot{m}_{in} - \sum \dot{m}_{out} = \Delta m_{sys}$$

A component mass balance on carbohydrate:

$$\dot{m}_{dr} \cdot X_{cdr} - \dot{m}_e \cdot X_{ce} = 0$$

Solve the above equation for X_{ce}:

$$X_{ce} = \frac{\dot{m}_{dr} \cdot X_{cdr}}{\dot{m}_e}$$

$$X_{ce} = \frac{2{,}435 \cdot 0.22}{3{,}851} = 0.14$$

The carbohydrate content in the extruder barrel is 14% wet basis.

3. Calculate the protein content

$$\sum \dot{m}_{in} - \sum \dot{m}_{out} = \Delta m_{sys}$$

A component mass balance on protein:

$$\dot{m}_{dr} \cdot X_{pdr} - \dot{m}_e \cdot X_{pe} = 0$$

Solve the above equation for X_{pe}:

$$X_{pe} = \frac{\dot{m}_{dr} \cdot X_{pdr}}{\dot{m}_e}$$

$$X_{pe} = \frac{2{,}435 \cdot 0.57}{3{,}851} = 0.36$$

The protein content in the extruder barrel is 36% wet basis.

4. Calculate the fat content

$$\sum \dot{m}_{in} - \sum \dot{m}_{out} = \Delta m_{sys}$$

A component mass balance on fat:

$$\dot{m}_{dr} \cdot X_{fdr} + \dot{m}_{fp} - \dot{m}_e \cdot X_{fe} = 0$$

Solve the above equation for X_{fe}:

$$X_{fc} = \frac{\dot{m}_{dr} \cdot X_{fdr} + \dot{m}_{fp}}{\dot{m}_e}$$

$$X_{fe} = \frac{2{,}435 \cdot 0.10 + 608}{3{,}851} = 0.22$$

The fat content in the extruder barrel is 22% wet basis.

5. Calculate the ash content

$$\sum \dot{m}_{in} - \sum \dot{m}_{out} = \Delta m_{sys}$$

A component mass balance on ash:

$$\dot{m}_{dr} \cdot X_{adr} - \dot{m}_e \cdot X_{ae} = 0$$

Solve the above equation for X_{ae}:

$$X_{ae} = \frac{\dot{m}_{dr} \cdot X_{adr}}{\dot{m}_e}$$

$$X_{ae} = \frac{2{,}435 \cdot 0.01}{3{,}851} = 0.006$$

The ash content in the extruder barrel is 0.6% wet basis.

III. Energy Balance Calculations
   A. Energy balance is:
   Using Q to indicate energy flow per unit time $\left(\text{such as } \dfrac{kJ}{hr}\right)$

   $$\sum Q_{in} - \sum Q_{out} + \sum \Delta H_{react} = \Delta h_{sys}$$

   where: $\sum Q_{in}$ = Sum of all energy flows into the system
   $\sum Q_{out}$ = Sum of all energy flow out of the system
   $\sum \Delta H_{react}$ = Sum of all energy released by reactions
   Δh_{sys} = Change in enthalpy of the system

   1. Energy into system includes:
      a) Steam injection (Q_s)
      b) Water injection (Q_w)
      c) Dry recipe (Q_{dr})
      d) Fat (lipid) injection (Q_f)
      e) Other additives
      f) Added through heated barrel jackets (Q_b)
      g) Added mechanically via viscous dissipation (Q_{me})
   2. Energy out of system includes:
      a) Steam lost at preconditioner/extruder transition (Q_{slp})
      b) Steam lost at vent (Q_{slv})
      c) Product exiting die (Q_{prod})
      d) Steam flashed at die (Q_{sld})
      e) Removed through cooled barrel jackets (Q_b)
      f) Lost by natural convection from preconditioner and extruder barrel surface (Q_c)
   3. Reaction energy in the system includes:
      a) Energy required for denaturation of protein (Q_{pd})
      b) Energy required for gelatinization of starch (Q_{sg})
      c) Energy required for phase changes
   4. Energy accumulated in the system:
      a) Normally is considered to be zero because we want to operate at steady state
      b) Must allow time for equilibration before taking data
      c) Must be regulated by control loops on all mass flows and barrel temperature which are necessary to ensure steady state
   B. Application of Energy Balances
   1. Calculating energy content of mass flows containing multiple components
      a) The general equation for the energy content of a stream is:
      Using C_p to indicate constant pressure heat capacity and T to indicate temperature

$$Q = \dot{m} \cdot C_p \cdot (T - T_{ref})$$

b) Calculating specific heat

$$C_p = 1.424 \cdot X_c + 1.549 \cdot X_p + 1.675 \cdot X_f$$
$$+ 0.837 \cdot X_a + 4.187 \cdot X_m$$

where: heat capacity of carbohydrate = $1.424 \dfrac{kJ}{kg°C}$

heat capacity of protein = $1.549 \dfrac{kJ}{kg°C}$

heat capacity of fat = $1.675 \dfrac{kJ}{kg°C}$

heat capacity of ash = $0.837 \dfrac{kJ}{kg°C}$

heat capacity of water = $4.187 \dfrac{kJ}{kg°C}$

2. Calculation of Energy Content of Water and Steam Flows
   a) Water flows
      Calculate the energy content using the equation shown above for streams containing dry materials. As indicated, the specific heat of water is 4.187 kJ/kg °C.
   b) Energy content of a steam stream is calculated using the equation: $Q = \dot{m} \cdot h$
      Enthalpy of steam, h, can best be found in a steam table. In most cases, we use 30 psi saturated steam for injection into the preconditioner (h = 2,721 kJ/kg) and 100 psi saturated steam for injection into the extruder barrel (h = 2,770 kJ/kg)
3. Calculation of Mechanical Energy Input
   a) Definitions
      i) Gross Mechanical Energy Input
         Total mechanical energy supplied by the drive motor to the gearbox/speed reduction system
      ii) Net Mechanical Energy Input
         That portion of gross mechanical energy input which is actually dissipated into the product—Gross Mechanical Energy Input minus losses in speed reduction system and bearings

b) Calculating mechanical energy input for a drive that operates at constant power with percent load indicator Using L to indicate percent load and P to indicate full load power: $Q_{me} = L_e \cdot P_e \cdot 36$

c) Calculating mechanical energy input drives that operate at constant torque up to the rated speed then at constant power above the rated speed (such as many DC drives):
Using N to indicate extruder shaft speed in rpm:
For $N_a < N_f$

$$Q_{me} = L_e \cdot P_e \cdot \left(\frac{N_a}{N_f}\right) \cdot 36$$

For $N_a > N_f$

$$Q_{me} = L_e \cdot P_e \cdot 36$$

To calculate net mechanical energy input, subtract the percent load when operating with no material in the barrel from the percent load when producing the product.

d) For three-phase motors:
Using V to indicate applied voltage and I to indicate current drawn: $Q_{me} = 3.6 \cdot \sqrt{3} \cdot V \cdot I \cdot \cos \phi$
Power Factor ($\cos \phi$) is commonly estimated at 0.8–0.85
To calculate net mechanical energy input, subtract the operating current when operating with no material in the barrel from the operating current when producing product.

4. Calculation of reaction energy, the energy consumed or evolved from a chemical reaction or phase change
   a) Starch gelatinization

$$Q_{sg} = \dot{m} \cdot X_c \cdot (X_{sg_{out}} - X_{sg_{in}}) \cdot \Delta H^\circ_{sg}$$

$$-10 \, \frac{kJ}{kg} \geq \Delta H^\circ_{sg} \geq -19 \, \frac{kJ}{kg}$$

Usually use: $\Delta H^\circ_{sg} = -14 \, \frac{kJ}{kg}$

   b) Protein denaturization

$$Q_{pd} = \dot{m} \cdot X_p \cdot (X_{pd_{out}} - X_{pd_{in}}) \cdot \Delta H^\circ_{pd}$$

$$-90 \frac{kJ}{kg} \geq \Delta H^\circ_{pd} \geq -100 \frac{kJ}{kg}$$

Usually use: $\delta H^\circ_{pd} = -100 \dfrac{kJ}{kg}$

C. Energy Analysis of Salmon Feed Production (Calculating the temperature of the product just behind the die in the process described in the figure above.)
   1. Calculate the energy associated with mass flows
      a) Incoming dry recipe
         i) First calculate the specific heat of this stream:

$$C_{pdr} = 1.424 \cdot X_{cdr} + 1.549 \cdot X_{pdr} + 1.675 \cdot X_{fdr} + 0.837 \cdot X_{adr} + 4.187 \cdot X_{mdr}$$

$$C_{pdr} = 1.424 \cdot 0.22 + 1.549 \cdot 0.57 + 1.675 \cdot 0.10 + 0.837 \cdot 0.01 + 4.187 \cdot 0.10$$

$$C_{pdr} = 1.79 \left(\frac{kJ}{kg\,^\circ C} \right)$$

   ii) Now, using the specific heat, calculate the energy content:

$$Q_{dr} = \dot{m}_{dr} \cdot C_{pdr} \cdot (T_{dr} - T_{ref})$$
$$Q_{dr} = 2{,}435 \cdot 1.79 \cdot (25 - 0)$$
$$Q_{dr} = 108{,}966 \frac{kJ}{hr}$$

   b) Water injected into preconditioner
$$Q_{wp} = \dot{m}_{wp} \cdot C_{pw} \cdot (T_{wp} - T_{ref})$$
$$Q_{wp} = 346 \cdot 4.187 \cdot (65 - 0)$$
$$Q_{wp} = 94166 \frac{kJ}{hr}$$

   c) Steam injected into preconditioner
$$Q_{sp} = \dot{m}_{sp} \cdot h_{sp}$$
$$Q_{sp} = 199 \cdot 2{,}721$$
$$Q_{sp} = 541{,}479 \frac{kJ}{hr}$$

   d) Fat injected into preconditioner
$$Q_{fp} = \dot{m}_{fp} \cdot C_{pf} \cdot (T_{fp} - T_{ref})$$
$$Q_{fp} = 608 \cdot 1.675 \cdot (50 - 0)$$

$$Q_{fp} = 50{,}920 \ \frac{kJ}{hr}$$

e) Water injected into extruder
$$Q_{we} = \dot{m}_{we} \cdot C_{pw} \cdot (T_{we} - T_{ref})$$
$$Q_{we} = 822 \cdot 4.187 \cdot (65 - 0)$$
$$Q_{we} = 22{,}317 \ \frac{kJ}{hr}$$

f) Steam injected into extruder
$$Q_{se} = \dot{m}_{se} \cdot h_{se}$$
$$Q_{se} = 181 \cdot 2{,}770$$
$$Q_{se} = 501{,}370 \ \frac{kJ}{hr}$$

2. Calculate the mechanical energy input for the drive operating at constant power with a percent load indicator:
$$Q_{me} = L_e \cdot P_e \cdot 36$$
$$Q_{me} = 34 \cdot 112 \cdot 36$$
$$Q_{me} = 137{,}088 \ \left(\frac{kJ}{hr}\right)$$

3. Calculate the reaction energy for the process
   a) Starch gelatinization
$$Q_{sg} = \dot{m}_e \cdot X_{ce} \cdot (X_{sg_{out}} - X_{sg_{in}}) \cdot \Delta H^\circ_{sg}$$
$$Q_{sg} = 3{,}851 \cdot 0.14 \cdot (0.90 - 0.17) \cdot 14$$
$$Q_{sg} = 5{,}510 \ \frac{kJ}{hr}$$

   b) Protein denaturation
$$Q_{pd} = \dot{m}_e \cdot X_{pe} \cdot (X_{pd_{out}} - X_{pd_{in}}) \cdot \Delta H^\circ_{pd}$$
$$Q_{pd} = 3{,}851 \cdot 0.36 \cdot (0.95 - 0.50 \cdot 95$$
$$Q_{pd} = 59{,}267 \ \frac{kJ}{hr}$$

4. Calculate the energy balance
$$\sum Q_{in} - \sum Q_{out} + \sum \Delta H_{react} = \Delta h_{sys}$$
$$Q_{dr} + Q_{wp} + Q_{sp} + Q_{fp} + Q_{we} + Q_{se} + Q_{me} - Q_{sg} - Q_{pd} - Q_e = 0$$
$$Q_e = Q_{dr} + Q_{wp} + Q_{sp} + Q_{fp} + Q_{we} + Q_{se} + Q_{me} - Q_{sg} - Q_{pd}$$

$$Q_e = 108{,}966 + 94{,}166 + 541{,}479 + 50{,}920 + 22{,}317$$
$$+ 501{,}370 + 137{,}088 - 5{,}510 - 59{,}267$$

$$Q_e = 1{,}391{,}529 \ \frac{kJ}{hr}$$

5. Calculate the temperature

$$Q_e = \dot{m}_e \cdot C_{pe} \cdot (T_e - T_{ref})$$

Solve for T_e:

$$T_e = \frac{Q_e}{\dot{m}_e \cdot C_{pe}} + T_{ref}$$

First, we need to calculate the specific heat:

$$C_{pe} = 1.424 \cdot X_{ce} + 1.549 \cdot X_{pe} + 1.675 \cdot X_{fe}$$
$$+ 0.837 \cdot X_{ae} + 4.187 \cdot X_{me}$$

$$C_{pe} = 1.424 \cdot 0.14 + 1.549 \cdot 0.36 + 1.675 \cdot 0.22$$
$$+ 0.837 \cdot 0.006 + 4.187 \cdot 0.27$$

$$C_{pe} = 2.26 \left(\frac{kJ}{kg°C} \right)$$

Now, calculate T_e:

$$T_e = \frac{1{,}391{,}529}{3{,}851 \cdot 2.26} + 0$$

$$T_e = 160°C$$

D. Observations of the Energy Analysis
  1. Where does the energy input come from?
     a) Dry recipe: 7%
     b) Water injected into the preconditioner: 6%
     c) Steam injected into the preconditioner: 37%
     d) Fat injected into the preconditioner: 4%
     e) Water injected into the extruder barrel: 2%
     f) Steam injected into the extruder barrel: 35%
     g) Mechanical energy from the main drive motor: 9%
  2. If no preconditioner was used:
     a) Need to put this 37% of the energy into the process somehow
     b) Steam injection into the extruder barrel is already at maximum
     c) Only option is to use mechanical energy
        i) To replace the steam energy with mechanical energy will require a 135% load on the extruder drive motor

ii) Obviously, we will need a larger motor
iii) This will result in increased wear
iv) Product will probably not perform the same
3. Where does the energy input go?
   a) Starch gelatinization: 0.38%
   b) Protein denaturization: 4%
   c) Energy losses: if calculated, will amount to less than 1%
   d) Heating the product: 96%

NOMENCLATURE

$\cos\phi$ = Power factor

C_p = Heat capacity $\left(\dfrac{kJ}{kg°C}\right)$

C_{pdr} = Heat capacity of dry recipe $\left(\dfrac{kJ}{kg°C}\right)$

C_{pe} = Heat capacity of material in extruder barrel $\left(\dfrac{kJ}{kg°C}\right)$

C_{pf} = Heat capacity of fat = $1.675 \dfrac{kJ}{kg°C}$

C_{pw} = Heat capacity of water = $4.187 \dfrac{kJ}{kg°C}$

h = Enthalpy $\left(\dfrac{kJ}{kg}\right)$

h_{se} = Enthalpy of steam injected into extruder barrel $\left(\dfrac{kJ}{kg}\right)$

h_{sp} = Enthalpy of steam injected into preconditioner $\left(\dfrac{kJ}{kg}\right)$

ΔH°_{pd} = Heat of reaction for protein denaturization $\left(\dfrac{kJ}{kg}\right)$

ΔH°_{sg} = Heat of reaction for starch gelatinization $\left(\dfrac{kJ}{kg}\right)$

I = Operating electrical current (Amps)

L_e = Indicated load on drive motor (%)

$\dot{m}$ = Mass flow rate (kg/hr)

$\dot{m}_{dr}$ = Mass flow rate of dry recipe to system $\left(\dfrac{kg}{hr}\right)$

$\dot{m}_e$ = Total mass flow rate of material in extruder $\left(\dfrac{kg}{hr}\right)$

$\dot{m}_{fe}$ = Mass flow rate of fat to the extruder $\left(\dfrac{kg}{hr}\right)$

$\dot{m}_{fp}$ = Mass flow rate of fat to the preconditioner $\left(\dfrac{kg}{hr}\right)$

$\dot{m}_{se}$ = Mass flow rate of steam to the extruder $\left(\dfrac{kg}{hr}\right)$

$\dot{m}_{slp}$ = Mass flow rate of steam lost from the preconditioner $\left(\dfrac{kg}{hr}\right)$

$\dot{m}_{sp}$ = Mass flow rate of steam to the preconditioner $\left(\dfrac{kg}{hr}\right)$

$\dot{m}_{we}$ = Mass flow rate of water to the extruder $\left(\dfrac{kg}{hr}\right)$

$\dot{m}_{wp}$ = Mass flow of water to the preconditioner $\left(\dfrac{kg}{hr}\right)$

N_a = Actual operating screw speed (rpm)

N_f = Full (rated) screw speed (rpm)

P_e = Rated power of extruder drive motor (kW)

Q = Energy rate $\left(\dfrac{kJ}{hr}\right)$

Q_{dr} = Energy rate carried by dry recipe $\left(\dfrac{kJ}{hr}\right)$

Q_{fp} = Energy rate carried by fat injected into the preconditioner $\left(\dfrac{kJ}{hr}\right)$

Q_{me} = Mechanical energy rate $\left(\dfrac{kJ}{hr}\right)$

Q_{pd} = Energy rate consumed by protein denaturation $\left(\dfrac{kJ}{hr}\right)$

Q_{se} = Energy rate carried by steam injected into the extruder barrel $\left(\dfrac{kJ}{hr}\right)$

Q_{sg} = Energy rate consumed by starch gelatinization $\left(\dfrac{kJ}{hr}\right)$

Q_{sp} = Energy rate carried by steam injected into the preconditioner $\left(\dfrac{kJ}{hr}\right)$

Q_{we} = Energy rate carried by water injected into the extruder barrel $\left(\dfrac{kJ}{hr}\right)$

Q_{wp} = Energy rate carried by water injected into the preconditioner $\left(\dfrac{kJ}{hr}\right)$

T = Temperature (°C)
T_{dr} = Temperature of dry recipe (°C)
T_{fp} = Temperature of fat injected into preconditioner (°C)
T_e = Temperature of product in extruder barrel just behind the die (°C)
T_{ref} = Reference temperature (°C)
T_{we} = Temperature of water injected into extruder barrel (°C)
T_{wp} = Temperature of water injected into the preconditioner (°C)
V = Operating electrical voltage (Volts)
X_a = Mass fraction of ash
X_{adr} = Mass fraction of ash in dry recipe
X_{ae} = Mass fraction of ash in extruder barrel
X_c = Mass fraction of carbohydrate
X_{cdr} = Mass fraction of carbohydrate in dry recipe
X_{ce} = Mass fraction of carbohydrate in extruder barrel
X_f = Mass fraction of fat

X_{fdr} = Mass fraction of fat in dry recipe
X_{fe} = Mass fraction of fat in extruder barrel
X_m = Mass fraction of moisture
X_{mdr} = Mass fraction of moisture in dry recipe
X_{me} = Mass fraction of moisture in extruder barrel
X_p = Mass fraction of protein
X_{pdr} = Mass fraction of protein in dry recipe
X_{pe} = Mass fraction of protein in extruder barrel
$X_{pd_{in}}$ = Mass fraction of denatured protein in
$X_{pd_{out}}$ = Mass fraction of denatured protein out
$X_{sg_{in}}$ = Mass fraction of gelatinized starch in
$X_{sg_{out}}$ = Mass fraction of gelatinized starch out

USEFUL CONVERSION FACTORS

$$\frac{\text{lbs}}{\text{hr}} \times 0.454 = \frac{\text{kg}}{\text{hr}}$$

$$\frac{\text{lbs}}{\text{min}} \times 27.24 = \frac{\text{kg}}{\text{hr}}$$

$$\text{horsepower} \times 0.746 = \text{kW}$$

$$\text{kW} \times 3{,}600 = \frac{\text{kJ}}{\text{hr}}$$

$$\frac{°F - 32}{1.8} = °C$$

$$(°C \times 1.8) + 32 = °F$$

$$\frac{\text{Btu}}{\text{lb°F}} \times 4.183 = \frac{\text{kJ}}{\text{kg°C}}$$

$$\frac{\text{Btu}}{\text{lb}} \times 4.183 = \frac{\text{kJ}}{\text{kg°C}}$$

Index

Adjustable jaw chokes, 75
Advantages of extrusion, 3, 83
After ripening, 27
Agglomeration, 2
Alkylresorcinols, 141
Allergen, 56
Amaranth, 184
Amino acid digestibility, 59
Ancient grain, 184
Anthocyanin pigments, 139
Antioxidants, 136, 137
Application of dry extruders, 53
Arabinose, 132
Ascorbic acid, 137
Atmospheric preconditioner, 119, 123, 89
Autogenous extruders, 10, 85
Automation, 3
Average retention time, 123

Baking soda, 201
Barley starch, 200
Barrel design, 96–97
Barrel wall, 98–99
Barrel wear, 102
Beads, 189
Beater configuration, 124
Benefits of preconditioned, 117

Beta-carotene, 137
Beta-glucans, 132
BHT, 137
Bile acid, 133
Bin, 28, 29
Bleaching agent, 180
Bowl life, 189
Bulk density, 35, 71

Calcium chloride, 180
Calcium hydroxide, 139
Canola, 73
Carotenoides, 137
Chemical changes, 128
Chlorprophan, 141
Choking, 65
Cholic acid, 132
Classification of dry extruder, 52
Co-extrusion dies, 91, 104
Collets, 73
Common uses of single screw extruders 170
Common uses of twin screw extruders, 170
Conduction, 40
Configuration, 31
Conical elements, 103–104

Index

Convection, 40
Conversion energy, 39
Cooking zone, 33, 40, 93–94
Corn brain, 132
Corn cones, 185
Corn flour, 185–186
Corn meal, 185
Critical parameters, 109
Cutter assemblies, 69
Cutting device, 91

Definition of extrusion, 1
Degassing, 2
Degerminated corn meal (softer texture) 43, 196
Degerminated corn meal (crunchier texture), 43, 196
Degree of fill, 124
Dehydration, 2
Deoxycholic acid, 132
Dependent variable, 37, 109
Design limits, 106
Dextrin, 71, 130
Dextrinization, 119
Dialyzable iron, 138
Die, 81, 88, 90
Die adopter, 104
Die plate, 35
Dietary Fiber, 131
Digestibility, 3
Direct expanded breakfast cereals, 183, 189
Direct expanded rings, 186
Direct expanded snacks, 170, 193–195
Dog food, 69–70
Double shaft preconditioned, 120
Dry extruder, 51

Elastometer, 82
Enzymes digestibility, 129
Expansion, 2
Extruded snack market, 83
Extruder assembly, 87–88, 90
Extruder barrel, 6, 31
Extruder capacity, 118
Extruder classification, 85
Extrusion cooking, 82
Extruder derive, 30
Extruder development, 4
Extruder solution, 44–45
Extrusion-expelling, 58

Fat soluble vitamin, 137
Feed grind, 70
Feed stock, 5
Feeding device, 29
Feeding zone, 92
Ferrous sulfate heptahydrates, 139
Fiber content, 177
Fish eggs, 186
Fish feed, 70
Flaked cereals, 183
Flaking grits, 185
Flaking rolls, 189
Flatulence, 141
Flavors, 142
Flight, 6
Food allergens, 140
Food extruder, 2, 26
Food industry, 167
Forming zone, 202
Fortification, 139
Free fatty acids, 136
Friction, 82
Frying, 203
Full Fat soy, 72
Function of extruder, 2
Function of single screw extruder, 168
Function of twin screw extruder, 168

Gearbox, 94
Gear drives, 30
Gelatinization, 2, 71, 129, 188
Genistein, 139
Glass transition, 37
Glucose, 130
Glucosinolates, 141
Glycoalkaloids, 141
Gravimetric feeding, 30
Grinding, 2
Gun puffer, 182

Half products, 196
Heavy bran, 187
Helical screw, 81
Hemagglutins, 56, 141
Hemi cellulose, 132
High-amylo pectin starch, 129
High shear stress products, 39
High-temperature short time, 3, 167
High-temperature short time extruders, 86
History of extruder, 4–5, 168

Holding bin, 87
Homogenization, 2
Hopper, 29

Ileal digestibility, 58
Independent variable, 35, 108
Indirect expanded RTE breakfast cereals, 182
Ingredient selection, 27
Interrupted flight extruder, 63
Isoflavones, 56, 139
Isothermal extruder, 10, 85

Kamut, 184
Kibbled, 66
Kneading elements, 99–100
Kneading zone, 31, 35, 40, 92–93

Land thickness, 71
Leakage flow, 31
Lectin, 141
Legnin, 132
Lipase, 56, 72–73
Lipids, 135, 111
Lipid oxidation, 136
Lipoxygenase, 56
Longitudinal grooves, 34
Loss-in-weight feed, 30, 88
Low-cost extruder, 5, 52
Low-shear stress extruder, 35
Low shear stress product, 48
Low temperature extrusion, 49
Lysine, 134, 135
Lysinoalanines, 180

Maillard reaction, 136, 142
Mannose, 132
Masa, 198
Material and energy balance, 112
Meat analogs, 135, 172, 180
Meat extender, 171
Meat substitute, 181
Medium shear stress extruder, 35
Medium shear stress products, 46
Melt transition, 116
Metering device, 29, 87–88
Millet, 184
Mineral, 138
Mineral absorption, 138
Mixing, 2
Moisture, 111

Molecular cross linkage, 172
Monoglycerides, 198
Multi dimensional snacks, 202
Multi screw, 82

Natural toxin, 140
Near infra red reflectance—spectroscopy, 129
New generation extruder, 5
Niacin, 137
Nitrogen solubility index, 177

Oat flour, 188
Oat ring, 188
Oat starch, 200
Oligosaccharides, 130, 141
Operating parameter, 38

Particle size, 177
Pasteurization, 2
Pectin, 132
Phenolic compound, 139
Phytate, 138
Phytic acid, 56
Phytochemicals, 139
Phytoestrogen, 139
Pigment losses, 137
Piston, 81
Polymer industries, 82
Polytrophic extruder, 10, 85
Potato peel, 132, 139
Potato starch, 200–201
Preconditioner, 5, 87–88, 115, 178, 188–189
Preconditioner categories, 119, 120
Preconditioning hardware, 119
Preconditioner inlet, 122
Preconditioner operation, 122
Preconditioning, 78, 115
Pregelled starch, 201
Pressurized conditioning, 89
Pressurized preconditioning, 89, 120
Principle of drying extrusion, 51
Process conditions, 35
Processing conditions for single screw extruder 42
Processing zone, 91
Product characterization, 3, 109
Pro-oxidant, 136, 137
Processing variables, 108
Protein, 134
Protein denaturization, 2

Index

Protein dispersibility index, 177
Protein solubility, 134
Psyllium, 184

Raffinose, 141
Rancidity, 136
Rapid viscoanalyzer, 129
Raw material, 26
Raw material for RTE breakfast cereal, 184
Raw material for TVP, 174
Raw material specification for TVP, 176
Recycling by-products, 59
Refined bran, 187
Residence time, 70
Resistant starch, 131
Retention time, 123
Reverse pitch, 101
Riboflavone, 137
Rice bran, 65
Rice starch, 200
Rotating screw, 81
RTE breakfast cereal, 181–193

Salt, 201
Samolina, 48
Screw, 5, 81, 87, 102
Screw design, 98
Screw pressing, 64, 66
Screw speed, 112
Seasoning recipe, 46
Second generation extruder, 5
Segmented screw, 74
Selection of hardware, 27
Shaping, 2
Shear, 2
Shear lock, 99–100
Sheeting die, 91
Single flight screw, 98
Single screw extruder, 25, 82–85, 169, 178
Single screw extruder classification, 36
Single shaft preconditioned, 120
Slotted wall expander, 74
Slow speed extruder, 35
Smooth bore, 34
Snack extruders, 39
Sodium alginate, 180
Sodium bicarbonates, 188
Sodium chloride, 180
Solvent extraction, 73
Soybean composition, 176
Soybean extrusion, 55, 57–58

Soy lecithin, 180
Soy protein, 135
Spiral grooves, 34
Sprout inhibitor, 141
Stachyose, 141
Starch, 129
Starch digestibility, 131
Starch transformation, 118
Sugar beat pulp fiber, 132
Sulfur, 180
Supplier of extruders, 169
Synthetic rubber, 76

Tapioca starch, 201
Terminology, 5
Texture, 2
Textured vegetable protein, 171–181
Textured Vegetable Protein additives, 179
Texturized soy isolates, 135
Thermoplastic, 82
Thermoset, 82
Thiabendazole, 141
Thiamine, 137
Thiamine stability, 137
Third generation snacks, 170, 193, 196, 198
Thrust load, 94
Tocopherol, 137
Toxic compound, 4
Trans beta carotene, 137
Transglucosidation, 130
Triglycerides, 136
Troubleshooting guide, 44
Trypson inhibitor, 55, 56, 72, 140
Twin screw extruder, 81–84, 169
Twin screw extruder classification, 85
Twin screw drive design, 94

Urease activity, 56, 72, 141
Uronic acid, 132

V-belt, 30, 94, 96
Vacuum, 49
Vaporization heat, 67
Vent, 48
Vented barrel, 6
Vented head, 47
Vibratory feeder, 30
Vitamin, 137
Vitamin A, 137
Vitamin C, 137
Vitamin D, 137

Vitamin E, 137
Vitamin K, 137
Vitamin retention, 137

Water soluble vitamin, 137
Waxy corn starch (high-amylopectin), 129
Weaning foods, 131, 135
Wheat bran, 136
Wheat gluten, 174

Wheat starch, 200
White rice flour, 187

X-ray differential patterns, 132
Xylose, 132

Yellow corn cones, 185

Zinc, 139